C0-ASP-049

Embassies and Illusions

Harvard East Asian Monographs 113

John E. Wills, Jr.

Distributed by the HARVARD UNIVERSITY PRESS, Cambridge (Massachusetts) and London
1984

Embassies and Illusions

Dutch and Portuguese Envoys to K'ang-hsi, 1666–1687

Published by the COUNCIL ON EAST ASIAN STUDIES, HARVARD UNIVERSITY

The Council on East Asian Studies at Harvard University publishes a monograph series and, through the Fairbank Center for East Asian Research and the Japan Institute, administers research projects designed to further scholarly understanding of China, Japan, Korea, Vietnam, Inner Asia, and adjacent areas.

Publication of this volume has been assisted by a grant from The National Endowment for the Humanities. The findings and conclusions herein do not necessarily represent the view of the Endowment.

Library of Congress Cataloging in Publication Data

Wills, John E. (John Elliot), 1936-
Embassies and illusions.

(Harvard East Asian Monographs ; 113)
Bibliography: p.
Includes index.
1. Netherlands—Foreign relations—China. 2. China—Foreign relations—Netherlands. 3. Portugal—Foreign relations—China. 4. China—Foreign relations—Portugal. 5. China—Foreign relations—To 1912. I. Title. II. Series.
DJ149.C5W55 1984 327.510469 84-9505
ISBN 0-674-24776-0

To Connie, with love

Contents

Photographs on the dust jacket and page 121 are from Olfert Dapper, *Gedenkwaerdig Bedryf der Nederlandsche Oost-Indische Maetschappye op de Kuste en in het Keizerrijk van Taising of Sina* (Amsterdam, Jacob van Meurs, 1671). English translation by John Ogilby, *Atlas Chinensis. . . . by Arnoldus Montanus* (London, 1671). By permission of the Houghton Library, Harvard University.

Photographs are based on drawings done on the Van Hoorn Embassy.
Jacket—Reception of Imperial Presents
Page 121—Imperial Banquet

Acknowledgments

It is most fitting that in work on a series of such improbable and contingent diplomatic connections I have engaged in a good deal of cultural diplomacy of my own, occasionally mildly exasperating but never as completely frustrating as the experiences of my seventeenth-century ambassadors, and have benefited from the help, advice, and encouragement of scholars and archivists in six countries. My most important scholarly debts are to John King Fairbank, father of us all in worrying about the tribute system, who encouraged my work in this area since its beginning in 1959, and to Charles Ralph Boxer, who has been generous with encouragement and bibliographic advice in Portuguese and Dutch colonial history. Jonathan D. Spence and John L. Cranmer-Byng read part or all of the draft at various stages and made useful comments. I profited from many hours of discussion of Ch'ing state ceremonies with Christian Jochim. Others who have helped me to find sources, correct translations, and think through various points are: Celeste Anderson, Bobby Chamberlain, Joseph Chen, Dominic C.N. Cheung, Katharine S. Diehl, Bailey W. Diffie, Joseph Fletcher, Fu Lo-shu, James Lee, James J.Y. Liu, Lee Reams, Francis M. Rogers, Francis A. Rouleau, S.J., Shan Shih-yuan, George B. Souza, Teotonio R. de Souza, S.J., Ts'ui Chien-chün, Frederic Wakeman, John W. Witek, S.J., and Herman Wong.

I also am extremely grateful to the authorities of a number of archives for opening their treasures to me, arranging for microfilming or transcription, and helping me with problems of source-location and paleography. My greatest debt is, as usual, to the Algemeen Rijksarchief, The Hague. The hospitality and helpfulness of the staff of the Archivum Romanum Societatis Jesu are truly remarkable. I also am very grateful to Francis A. Rouleau, S.J., for loaning me microfilms from this archive. Also important for this study were the Historical Archives, Panjim, Goa, India, the Biblioteca da Ajuda and the Arquivo Histórico Ultramarino, Lisbon, and the Houghton Library of Harvard University. Scraps of information also were drawn from the India Office Library and Records, London, and the Arsip Nasional Republik Indonesia, Jakarta. If I do not list names of all those who helped me, it is not because I do not remember many of them by name and with gratitude, but because one is always aware in an archive how much depends on the efforts of devoted staff people whose names one never knows.

At various points in its ridiculously long evolution, the work leading to this book has been supported by a Sheldon Traveling Fellowship from Harvard University, small grants from the John H. and Dora B. Haynes Foundation of Los Angeles, from the National Defense Education Act Research and Publication Fund of the University of Southern California, and from the American Philosophical Society, and major fellowships from the National Endowment for the Humanities and from the American Council of Learned Societies. The Committee on Scholarly Communication with the People's Republic of China and the Chinese Academy of Social Sciences made it possible for me to see the palaces of Peking and clear up some puzzles about the location of embassy ceremonies in the summer of 1979. I am very grateful to all of these agencies for their generous support.

At the University of Southern California, the efficient help of the Interlibrary Loans staff of the University Library was crucial to this work. Clara Harada, JoEllen Pope, Vivian Smith, and others provided efficient typing services. Florence Trefethen of the Council on East Asian Studies at Harvard was a most careful and helpful editor.

For permission to publish a translation of the *Breve Relação* of Francisco Pimentel, S.J. (Appendix A), I am grateful to C. R. Boxer and to the authorities of the Houghton Library, Harvard University, and of the Biblioteca da Ajuda, Lisbon. For permission to publish and translate the passage from the Jesuit annual letter of 1678-1679 (Appendix B), I wish to thank Edmond Lamalle, S.J., Archivist of the Archivum Romanum Societatis Jesu.

My parents and Robert H. Irrmann have provided encouragement along the way and crucial financial assistance for the first year of work in The Hague, for which I remain grateful. My wife read and proofread this book at various stages, helped with it and had her life broken up by the travels it entailed. This book is dedicated to her in gratitude for these and far less tangible and more important kinds of support.

THE IMPERIAL PALACE, PEKING

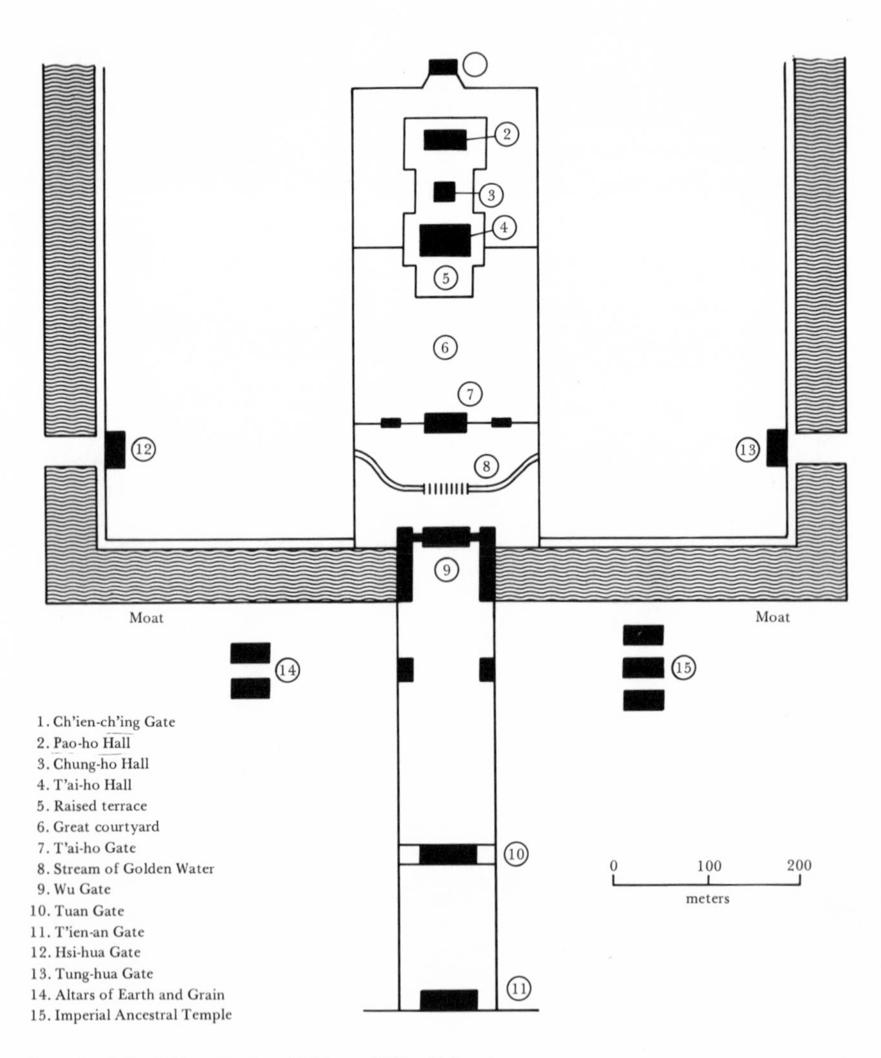

From plans in Osvald Siren, *The Imperial Palaces of Peking,* Volume 1.

chapter one

Continuities and Routines

THE AUDIENCE

The Honorable Pieter van Hoorn, Ambassador of Their Excellencies the Governor-General and Council of the United Dutch East India Company, was summoned from his lodgings with his suite long before dawn. At the T'ien-an Gate, they dismounted from their horses and proceeded on foot, through the tunnel over 20 meters long that pierces the massive gate structure, up the long avenue and through another huge gate, and on up into the cobbled courtyard 75 meters on a side, enclosed by the thrusting wings of the Wu Gate. Here they waited with a great crowd of officials in court dress, their ranks indicated by the animals embroidered on their robes and the buttons on their hats, who gawked at the foreigners as people no longer do in the capital but still do in Chinese country towns. The light revealed the deep, earthy red of the walls, the brightly painted designs under the eaves, and last of all the yellow of the roof tiles. The gate structure rose over 25 meters above the tunnels through it, and even the tunnel entrances dwarfed the imperial elephants and carriages that stood between them.[1]

When it was fully light, the officials lined up according to rank, the Ambassador and his suite taking their places at the end of the west, or military, line. They filed through the side tunnels, across the elegant marble bridges over the curving water channel, in the court between the Wu Gate and the fine, complex composition of the T'ai-ho Gate, the two side gates, and the corner towers. They ascended to the west side gate, passed through it, and descended into the court facing the T'ai-ho Hall. The modern visitor, deceived by the long, low proportions of each of the buildings and of the whole composition, has a hard time realizing that the court is 200 meters square, the peak of the roof of the Hall 35 meters above the pavement. The envoy and his suite probably grasped the scale more quickly, seeing the court full of officials seated in orderly rows, the central avenue leading to the Hall bordered with elaborate flags, umbrellas, and other imperial regalia. After a half hour of waiting, the beginning of the ceremony was announced by the sound of a great bell, then by the cracking of large whips that sounded like firecrackers. The Emperor had taken his place on the throne in the shadows of the Hall, but was invisible from far back in the court. Various officials passed south of the imperial regalia into the central avenue, took the positions designated by markers for their various ranks, and to the commands of a herald knelt, bowed until their foreheads touched the stones three times, rose, knelt, touched their foreheads to the ground three more times, rose, and repeated the cycle a third time. Then the Ambassador and his suite were summoned to perform the same ritual, taking great care that their big plumed hats did not fall off. They took the place assigned to a rank, probably a low rank, in the Chinese hierarchy. Their calisthenics completed, they were taken up side steps into the side of the Hall, where they could have a cup of tea and see the Emperor on his throne. There was a chance that the Emperor would stop and speak to them as he left the throne at the end of the audience, but that would have been a sign of special favor, and the Van Hoorn embassy was being treated very coolly.

This was the famous ceremony which in some ways formed the core of Chinese formal diplomacy down to the 1860s. It was an oddly perfunctory appendage to the most matter-of-fact of imperial ceremonies, the ordinary audience, which could be held as often as every ten days, and was used primarily as an occasion for officials to give thanks to the Emperor for new appointments and honors. It had nothing of the theatrical reinforcement of a sacred mystique found in the reception of foreign envoys by Byzantine emperors or Muscovite tsars. The distances in the great court strengthened a sense of focus on relations among positions rather than persons, and of the rather offhand arrogance of a court that saw itself as the center of the world and actually ruled a quarter of mankind. In this audience ceremony, and especially in the famous "three kneelings and nine prostrations" of the kowtow, the ambassador, or even the very occasional foreign monarch, was assimilated to the status of a minister of the Son of Heaven. Before and after the audience there were the other ceremonies of an embassy in Peking—banquets, exchanges of presents and documents—that expressed other symbolic themes; enmeshed with the ceremonies in the capital and in the border zones through which embassies arrived and departed were a complex of policies and institutions for managing not just embassies but a wide spectrum, sometimes all, of relations with one foreign power or another.

For a Korean, Ryukyuan, or Vietnamese envoy, paying homage at the core of his cultural universe, or for an Inner Asian envoy, seeing a Manchu conqueror enthroned as lord of both China and Inner Asia, this was a very impressive ceremony, and many of its symbolic and institutional associations would have been clear. We do not really know what Van Hoorn was making of these extraordinary sights; his descriptions of the audience and of his whole stay in Peking are matter-of-fact. He may have been impressed; certainly he did not go away with a negative attitude toward everything Chinese, for eight years later he published a long didactic poem on Confucius. He had no prior experience of the ceremonies of oriental courts, but he had been ordered to conform to those of the Chinese court, and his own culture, while not going in for prostrations, was a deferential and ceremonious one. His masters were aware that the Ch'ing rulers might regard an embassy as bearing tribute, but saw no reason to worry about that, for in fact ambassadors were received in Peking from powerful and independent princes. The Dutch did not mind going through a few tiresome ceremonies, even giving substance to a few heathen illusions, in order to carry on important negotiations with great sovereigns and their officials. Van Hoorn soon would discover that the Chinese regimen for an embassy gave almost no opportunity for such negotiations, but on the day of his audience his illusion that there would be substantial discussions still was intact.

The officials responsible for the reception of the Van Hoorn embassy had reviewed their empire's relations with the Dutch over the previous decade, and had seen them, as the Dutch never did, as applications and modifications of bureaucratic regulation of all aspects of relations with all foreign countries. This was a system inherited from the Ming, and, if the officials had wanted to consult precedents for a particular aspect of the management of an embassy, they might have consulted the *Collected Ceremonies of Great Ming,* which gave some parts of the embassy routine a pedigree extending back to the third millennium B.C. Western writers have called this ceremonial and institutional complex the tribute system. I have argued elsewhere that the tribute system was not all of traditional Chinese foreign relations, and may not be the best key to a comprehensive understanding of these relations.[2] The Western literature on early Sino-Western relations may have given excessive emphasis to tribute embassies.[3] But the system certainly was an important determinant of Ch'ing policy toward the Dutch at the time of the Van Hoorn embassy, and attitudes and institutions associated with it did continue to influence Ch'ing policies until the nineteenth century.

The Van Hoorn embassy was the first of four from maritime Europeans in just two decades. This burst of activity came at the beginning of the dynasty, in the early years of the great reign that would really consolidate its power and its insitutions. These embassies had diverse and distinctly unsystematic origins in the shifts of Ch'ing court politics and coastal policies and in the experiences and goals of the Dutch and the Portuguese. In a sense they led nowhere; by the end of these twenty-one years, the connection between the tribute system and the management of maritime trade was tacitly broken, and there were no more maritime European embassies for commercial purposes for more than a century. But, because of these embassies, later Ch'ing statesmen could look back on an apparently substantial record of tribute-system management of relations with western Europe. Thus, in these crucial decades of Ch'ing consolidation, a double movement of revival and then tacit abandonment of the tribute system contributed to the drift, illusion, and lack of systematic policy that characterized Ch'ing relations with the West until the middle of the nineteenth century.

The following chapters give full accounts of these four embassies and of the European and Chinese contexts and backgrounds necessary to understand them. These accounts also can be read as case studies in the functioning of the institutions of the tribute embassy. More such case studies would be most welcome, of embassies from the Confucian monarchies of Korea, Ryukyu, and Japan, of the Mongol, Tibetan, and other Inner Asian powers,[4] and especially of the very rare cases in which we have sources giving the views and experiences of a non-Confucian Asian ambassador.[5] Although three of these four embassies have been studied previously in modern times,[6] there has been no study that made use of all the Western-language or all the Chinese sources on any one of them. I have found new sources on them in Goa, Jakarta, Lisbon, Rome, and The Hague, as well as scraps of additional printed Chinese material.[7] This study begins by outlining the contexts that led Ch'ing officials to view these embassies as parts of a very long tradition and of a routinized bureaucratic system. Other Chinese and European contexts will be studied in their places in the next four chapters. Finally we will try to see what we can learn from these strange events in all their contexts about the basic patterns of Ch'ing diplomacy and Sino-Western relations.

TRIBUTE: A SKETCH FOR A HISTORY

In the period we are dealing with, the Ch'ing rulers had not yet compiled their own basic statutes; the first edition of the *Collected Statutes of Great Ch'ing* (*Ta-Ch'ing hui-tien*) was completed in 1690. Thus, the most comprehensive sets of precedents for the management of tribute embassies available to them in these years were those contained in the *Collected Statutes of Great Ming* (*Ta-Ming hui-tien,* 1587) and the *Collected Ceremonies of Great Ming* (*Ta-Ming chi-li,* 1530).[8] These precedents were not followed in every particular, but the fundamentals were not changed, and above all the Ch'ing rulers would have gotten from them a strong sense of the tribute system as a comprehensive pattern for the management of foreign relations embodying some very long continuities. The *Collected Ceremonies,* a product of the extreme antiquarianism and preoccupation with ceremonial issues of the early Chia-ching period, prefaced the rules for each set of ceremonies with references and quotations reaching back in many cases to the mythical Sage Emperors. For some ceremonial complexes a long and substantial tradition could be shown; for example, the Altars of Earth and Grain (She-chi-t'an) and the Imperial Ancestral Temple (T'ai-miao) stood on the west and east of that long avenue from the T'ien-an Gate to the Wu Gate, precisely in the positions in relation to the Palace in which they were supposed to have stood in the early Chou period. But almost none of the ceremonies of a tribute embassy were given a pre-Han lineage, and only for the reception of presents from tributaries was the lineage at all substantial. That lineage was extended back to the reign of the Yellow Emperor, to whose court, it was said, the Southern I had come riding on white deer and bearing gifts of sweet wine. One of the most important bits of evidence for the continuity of tribute from the early Chou period, quoted twice in the *Collected Ceremonies,* in several Ming memorials on tribute embassies, and in an early Ch'ing official's report on the Dutch, was the *Hounds of Lü* section of the *Classic of Documents* (*Shu-ching, Lü-ao*).[9] This text, now thought to be of post-Chou origin, also focused on and took its name from presents brought by tributaries, insisting that the true ruler would not prize rare gifts in themselves, but solely as evidence that his virtue had attracted homage from afar. In the *Tribute of Yü* (*Yü-kung*) section of the *Documents,* local products of various regions are offered to a central sovereign as symbols of his sovereignty, and the barbarians form the outer rings of a tidily concentric world, ideas that later were both parts of the world view of the tribute system; but the detail is almost all about presents from within the Chinese cultural sphere, and there are only cryptic references to specific groups of non-Chinese people.[10] There is evidence, however, that both the "Huai barbarians" and the "semi-barbarian" state of Ch'u were considered tributaries of the Chou.[11] All in all, it seems that by far the strongest continuity in the handling of relations with non-Chinese that later statesmen could trace back to the idealized early Chou was that of the central symbolic importance of the presents brought by the foreigners, and that presents of exotic animals aroused especially strong and ambivalent fascination. The classical ceremonial compendia that were the foundation of such long and detailed continuities in many ceremonial practices had very little to say about relations with peoples of non-Chinese culture; the few allusions to them in the Ming compendia are to commentaries or to brief and ambiguous passages; the sections on "ceremonies for the reception of guests" are entirely about visits and meetings among rulers and envoys within the sphere of Chinese culture.[12] It is clear that Chinese statesmen and thinkers of late Chou times attached much importance to the distinction between Chinese and barbarians, but *Tso-chuan* texts make it clear that these distinctions were not consistently honored in practice: "barbarians" occasionally were full participants in meetings among rulers of states, including the solemn swearing of oaths; detailed rules for these meetings are given in the classical compendia but without mention of barbarian participation or Sino-barbarian distinctions. The great northern state of Chin had especially complex relations with "barbarian" peoples, including a good deal of intermarriage of ruling houses.[13]

We also have some evidence for the early development in Chin of ideas about "management" of non-Chinese peoples. According to the *Tso-chuan,* in 568 B.C., a minister of this state advised friendly relations with the barbarians to the north; in this way Chin's armies would not be overexerted, her people could till their fields in peace, Chin would not be distracted from balance-of-power politics among the Chinese states, and, since the barbarians were more fond of goods than of land, Chin could buy land from them.[14] There is not much in later records about buying land from the barbarians, but a preoccupation with internal Chinese politics, an agrarian-bureaucratic preference for peace, and a tendency to use Chinese wealth to buy not land but peace from the barbarians are the most persistent themes in Chinese Inner Asian policy until the seventeenth century A.D.

These were the roots of a defensive policy, aiming at the prevention of attack by foreigners, not at active domination over them —a tendency that would be strengthened by the emergence of the threat of nomadic mounted archers on the steppes in the Warring States period and the building, by the successor states of Chin and others, of the defense works that were linked under the Ch'in into the Great Wall. The early Han rulers continued this defensive tendency, dealing with the menace of the Hsiung-nu nomad power by a policy of *ho-ch'in,* "conciliation and blood ties," addressing the Hsiung-nu rulers as near-equals, sending Han princesses to marry them, and sending them large annual gifts of silks and other fine Chinese goods, in return for Hsiung-nu promises not to raid the frontiers. But, by about 170 B.C., the Hsiung-nu were raiding more and more and the famous scholar Chia I was attacking the *ho-ch'in* policy as a violation of the proper hierarchical order focused on a single Son of Heaven.[15]

There was a striking break with this defensive tendency in the reign of the Emperor Wu (140–87 B.C.), which was one of the greatest periods of territorial expansion in the history of China. This expansion took a number of forms, but the practices, ceremonies, and rhetoric of tribute played only marginal roles in them. As small states in Central Asia came under Han domination they were expected to send envoys to the imperial court, "call themselves ministers" (*ch'eng-ch'en*), and present gifts, of which the most famous were the "Heavenly Horses" of Ferghana. But Han domination of that area depended less on the manipulation of these "tributary" practices than on the holding of hostage princes at the imperial court and the active—and frequently cynical and brutal—interference of Han envoys and armies in the western Regions. Korea and Nan-Yueh (modern Kwangtung), areas in which elite refugees from the post-Ch'in chaos had intermarried with the local rulers, had been enfeoffed as vassals of the Han, not far in status from rulers of kingdoms within China, had grown in power and independence, and now found their territories conquered and brought not into a tributary relation but into the terminology and formal status of bureaucratically administered areas within China. In a fascinating debate in 133 B.C. between advocates of attacking the

Hsiung-nu and those who wanted to renew the *ho-ch'in* policy, neither side used tributary rhetoric or advocated offering them tributary status; this was done in 119 when a Han proposal that the Hsiung-nu ruler declare himself an "outer minister" (*wai-ch'en*) of the Han was angrily rejected and led to a period of all-out warfare.[16]

Another step toward the general application of tributary ceremony and rhetoric was taken in 53 B.C., when the ruler of one group of the now divided Hsiung-nu came to court and was received as a "border minister" (*fan-ch'en*). After considerable debate among the Han officials, he was received with the ceremonies appropriate for a guest, not a minister, in court ceremonies was given precedence over the feudal lords and kings within China, and his name was not announced; in that way, if he or his successors did not remain submissive, their lineage would not have been announced by name or received as ministers at the court, and the Han would not be honor-bound to treat them as rebels.[17] This is the first case I know of in which we can see Chinese officials using the ceremonial side of foreign relations to assert formal suzerainty while avoiding commitment to actual domination of areas outside China.

In the double movement of the rise of Wang Mang and the restoration of the Han, the ceremonial and rhetorical framework of foreign relations, like many aspects of state ceremonial, were first subjected to the most willful and dogmatic alterations and then restored to a clarified and systematized version of practices that had been developing under the early Han. Wang Mang aroused the antagonism of many non-Chinese rulers by calling them dukes, not kings, since "king" had been the title of the Chou ruler and "Heaven has not two suns nor Earth two kings." He even tried to change the name Hsiung-nu to Hsiang-nu, "surrendered slaves."[18] In the later Han period, many non-Chinese were settled within Chinese jurisdiction as "dependent countries" (*shu-kuo*), which apparently were dealt with as parts of the administrative structure of the empire, not as tributaries. All other foreign rulers who had dealings with the court were treated as tributaries. Tribute-bearers who were in the capital at the New Year participated in the spectacular New Year audience and banquet, with over 10,000 people present. It is not clear what ceremonies tribute-bearers performed at other times of year; perhaps they simply had to wait for the next New Year.[19]

The compilers of the *Ta Ming chi-li* had considerably more information about the T'ang reception of tribute envoys than they did for earlier periods—more than I have been able to find in standard histories and compilations—and for the first time provide information on the origins of the main ceremonies that still were important in the Ming era. The T'ang regimen began with an elaborate ceremony of "welcoming and rewarding for toil" (*ying-lao*) at the tributaries' official lodgings in the capital, followed by a "warning of [impending] audience" (*chieh-chien*), and then a single spectacular ceremony in which the envoys were received and banqueted and their presents displayed before the emperor. The tributaries kowtowed every time an imperial word was transmitted to them; when the emperor raised his wine cup for the first time, special music was sung and the tributaries kowtowed again. For lesser tributaries or for minor envoys even from important ones, there was less music, and fewer banners were displayed; but in general the ceremony seems to have been uniform for all.[20]

These ceremonies were the manifestation of a real will toward domination equaled in only a few periods of Chinese history. The early T'ang rulers demanded acknowledgment of their suzerainty from all foreign states having any relation with China, including the formidable steppe empires of the Turks and the Uighurs. They backed their demands with astute divide-and-rule politics toward the Turks, expeditions and garrisons in the Western Regions, and the last uses in Chinese history of the *ho-ch'in* policy toward both the Turks and the Uighurs. The steppe menace seldom was far from Ch'ang-an, but early T'ang emperors and generals, closer to the steppe in their own culture and style of leadership than Sung and Ming rulers, managed to face down invaders in personal confrontations or mollify them with especially lavish and ceremonious receptions at the capital.[21] In the perilous years during and just after the An Lu-shan rebellion, when the survival of the T'ang dynasty depended on Uighur support, the great general Kuo Tzu-i awed the Uighurs with his bravery and eloquence, and even wished an imperial "ten thousand years" (*wan-sui*) to both the T'ang Emperor and the Uighur Khaghan. The T'ang Crown Prince once saluted a Uighur general to persuade him not to plunder the capital. As the crisis passed, the T'ang continued to mollify the Uighurs by sending princesses to marry the khaghans and by paying fixed amounts of silk for each horse they brought to the capital in connection with their embassies, the first well-documented case of a kind of controlled trade that was very important under the Ming. They tried to limit the numbers of horses and people coming to the capital, sometimes were successful, and sometimes gave way to Uighur extortion.[22] But, unlike the Han, who got only peace in return for their silks and princesses, the T'ang got horses and acknowledgment of their suzerainty, used the Uighurs against the Tibetans, and survived until the breakup of the Uighur empire. In summary, the T'ang probably represented the summit of traditional Chinese diplomacy, controlling dangerous foreigners and demanding their ceremonial submission, using Chinese resources of wealth, bureaucratic coherence, and psychological astuteness, attentive to detail but flexible and realistic in handling embassies to the capital and sending T'ang envoys out to foreign rulers.

Sung ceremonies for the reception of foreign envoys followed T'ang precedents in many respects, but with an unusual amount of differentiation in the ceremonies and bureaucratic agencies involved; Liao envoys were the most favored, followed by those from Koryo and Hsi-hsia, then by those from less powerful and sophisticated polities.[23] These differentiations were signs of realistic recognition of some fundamental changes. The Sung faced a new kind of challenge in regimes—the Liao and Hsi-hsia—based on non-Chinese tribal organization but also incorporating Chinese officials and institutions, not to be satisfied with plunder and gifts, understanding that the real wealth was the right to tax agricultural land. The Treaty of Shan-yuan with the Liao in 1005 also has been viewed by some as an act of realistic statesmanship, not desperation, in which the Sung state adjusted to its changed situation and bought off the Liao for annual payments modest in comparison to potential defense expenses and to the growing revenues of the Sung state. There is much truth in this, but the *consequences* of this settlement contributed to some decidedly unrealistic trends in Chinese attitudes toward foreign relations.

Whereas the T'ang had been able to recover from foreign incursions and regain formal control of its territory, the Sung now had formally conceded sixteen prefectures in the northeast to the Liao. The acknowledgment of the parity of the Sung and Liao emperors was not a solitary gesture like Kuo Tzu-i's *wan-sui* but a feature of an ongoing diplomatic relationship, expressed in every exchange of envoys. Almost all Sung officials viewed the northerners as barbarians and accommodation with them as a temporary expedient at best. Combined with the permanent loss of territory and the constantly reiterated affront of ceremonial equality, these traditional attitudes produced a revanchist obsession that constantly affected Sung policy-making, contributed to some disastrous errors of judgment, and on the ceremonial side led the Sung rulers to resort to various petty shifts to avoid fully acknowledging Sung-Liao parity—sending low-ranking envoys to the Liao, placing Liao envoys in low-ranking positions at Sung audiences, and so on.[24] This tendency to concentrate on ceremonial appearances when the realities of power were obdurately unpleasant was reinforced by the growth of a Neo-Confucian political culture, most strikingly exemplified by Ou-yang Hsiu and his generation, that was intensely serious, we would say obsessive, about many matters of ceremonial form.[25] It was not balanced by the realism and attention to detail and historical experience seen in many other areas of Sung political discourse, I suspect in part because, although many Sung officials had been on one embassy abroad, few of them had as much realistic experience of foreign affairs as they had of land tax, labor service, water control, currency problems, and so on. Bursts of diplomatic activism in the Northern Sung came in periods when "reformers" and "post-reformers" were influential, but neither Wang An-shih nor Ts'ai Ching seems to have been much interested in foreign policy.

Sung focus on the forms of Chinese superiority was dysfunctional in dealing with the realities of inter-state politics. It may have contributed to the breakdown of a possible reconciliation with the Hsi-hsia in the 1060s.[26] It certainly contributed to the disastrous mishandling of relations with the rising Chin power after 1100. Here the Sung rulers faced a challenge that might have baffled the most astute masters of realpolitik, in the threat of the explosively expanding Jurchen Chin to overwhelm the Liao, including its territories south of the Great Wall, and thus pose a fresh and formidable threat to the Sung. The Sung ministers sought negotiations with the Chin in the hope that by a military alliance with the Chin against the Liao they could both crush the Liao and recover the lost territories south of the Wall, apparently never seeing that they could do both at the same time only with the acquiescence of the extremely energetic and capable Chin ruler with whom they sought an alliance. They delayed and complicated negotiations with petty struggles for ceremonial superiority, referred to the people they were negotiating with as "barbarians" (*lu*), and manifested a sublime faith in their own ability to get what they wanted by complex and tricky manipulation of the negotiations, when in fact it seems that the Chin Emperor was seeing through their every move and manipulating *them* with consummate skill.[27]

The Sung agreed in 1123 to deal with the Chin as equals, using the forms specified by the Treaty of Shan-yuan with the Liao, but after the Chin conquest of north China (1125–1127) some Chin envoys apparently were not received by the Emperor at all, while one was received only after strenuous efforts to bring the Southern Sung retinues and regalia up to Northern Sung standards.[28] The greatest contributor to a Chinese tradition of foreign relations in the Southern Sung was Yueh Fei,[29] and, while his condemnations focused on the disgrace of acquiescence in territorial loss rather than on the shame of acceptance of ceremonial equality, the two were linked in relations with the Liao and then the Chin, and the linkage between the moral-critical tradition of post-T'ang Confucianism and insistence on the ceremonial superiority of the Chinese emperor was forged more firmly than ever.

In this rather breathless survey of some aspects of the formation of a Chinese tradition of foreign relations down to the end of the Sung, we have seen Chinese officials paying much attention, some astute, some misguided, to ceremonial forms of embassies, and especially to maintaining the ceremonial supremacy of the Son of Heaven; efforts to manage trade in connection with embassies and especially to limit and control the size of this trade and the number of foreigners coming to the capital; the use of presents, trade concessions, titles, ceremonious receptions to mollify dangerous foreigners. There are a few indications of T'ang and Sung efforts to establish Chinese regulations on the frequency of embassies and the size of embassy parties. What we have *not* seen is the coalescence of all these tendencies into a single pattern for the bureaucratic management of all foreign relations which we might call a tribute *system*. That was the work of the early Ming. The new systematization was part of the general institution-building and bureaucratic systematization in the early Ming; the increased restrictions on foreign contact were a result of the systematization of the earlier trends and tendencies mentioned above, of successive critical changes during the Ming in the relations between the emperor and the bureaucratic elite, and of the new importance in foreign policy-making of maritime phenomena and issues.

When the Mongols withdrew from Peking in 1368, the north China plain had not been under a Chinese monarch since 1125, the Peking area not since 935. The moralistic political culture of Neo-Confucianism that had emerged in those centuries, linked to and intensified by culturalist reaction against a near-century of Mongol rule of all of China and by the actions of a ruler with an awesome will to systematize and settle everything for succeeding generations, produced in the early Ming a great burst of comprehensive and dogmatically uniform system-building. The establishment of a proper regimen of state ceremonies was very serious business indeed, both for the scholar-officials and for an Emperor anxious to buttress his own legitimation and extremely serious about the interconnections between his ceremonial acts and natural phenomena.[30] A great re-survey of land and re-assessment of taxes on it culminated in the presentation of the registers on the Altar of Heaven.[31] A uniform system of prefectural schools and state-supported scholars was set up, and

Chu Hsi's interpretations of the Classics were made the orthodox standard for schools and examinations.[32]

An integral part of this energetic early phase of Ming government was the reassertion of the ceremonial and institutional supremacy of the Son of Heaven over all other sovereigns. Envoys were dispatched in 1368 bearing edicts summoning foreign rulers to send envoys to present tribute to the new dynasty and to request confirmation of their positions from it. When the King of Annam came in person to pay homage in 1369, he was greeted by the most elaborate set of ceremonies ever used in the tribute system: three banquets before the audience; the King's presents and tribute memorial displayed at the audience, the King in person handing to the Emperor the memorial and a list of the presents; somewhat more modest audiences before the Heir Apparent and the other Princes of the blood; a great banquet in the presence of the Emperor and all the Princes, with elaborate specifications of music to accompany the successive stages and even a special dance, "Various Countries Come to Pay Court"; two more banquets at the court of the Heir Apparent and at the lodgings for tributaries; further full court audiences to take leave of the Emperor and the Heir Apparent. This routine apparently was used for most rulers and envoys coming to pay tribute during the Hung-wu period.[33] The Ming court also specified the ceremonies that were to be used at the court of a tributary state when envoys were sent off to the Ming court or when an edict or a seal granted by the emperor was received.[34]

Beyond the purely ceremonial sphere, the Hung-wu Emperor established a tally system for checking on the authenticity of embassies from Siam, and may have made a first effort to limit Annamese embassies to one every three years, the first cases of types of restrictions that would be applied to most tributaries in subsequent reigns.[35] But his most important innovation in foreign policy was to ban all trade in Chinese ports by foreigners not connected with tribute embassies and to forbid all Chinese voyages abroad, so that China's only legal maritime trade was that carried on within the framework of the tribute system.[36] This was a defensive reaction, characteristic of traditional Chinese foreign policy but in the long run extremely counterproductive in this case, to the attacks of Japanese pirates, joined in these early years by many partisans of two of the defeated rivals of the first Ming Emperor. Ming edicts threatening invasion of Japan if the pirates were not stopped elicited defiant responses, and relations finally were broken off completely in the 1380s. Coastal defenses were strengthened, and there were new efforts to cut off trade not connected with embassies. In the 1390s, there were repeated orders to punish heavily Chinese who were engaged in non-tributary trade; embassies from Southeast Asia were welcomed as a legal alternative to such trade, but there simply were not enough of them.[37]

In an interesting parallel to these tendencies to place restrictions on tribute embassies, the ceremonial for their reception was considerably simplified in 1394. The presents for the Emperor and the tribute memorial now were turned in to high officials, not displayed at the audience. No longer was every envoy given a banquet in the presence of the emperor; sometimes the emperor sent wine, which was received with elaborate ceremony, for a banquet at the envoys' lodgings. I think it likely that more ceremonies were eliminated from this complex in 1394 or later in the dynasty, along the way to the very much simpler complex of ceremonies we will see in the early Ch'ing. But the subject does not seem to have aroused great interest among Ming scholars and recorders of state ceremony, and our documentation is very limited.[38]

In the Yung-lo reign (1402-1424), Chinese envoys were sent out in all directions to gather information about foreign peoples and to encourage foreign rulers to send tribute embassies. The great maritime expeditions of Cheng Ho reached as far as East Africa, bringing back a Somali embassy with a giraffe for the Emperor—perhaps the high point of the "Hounds of Lü" exotic-animal continuity of the tribute tradition. Land missions reached across the Pamirs to Herat and Samarkand, across the Himalayas to Bengal, and to the Amur River in the northeast. The Emperor personally led military expeditions across the Gobi to break continued Mongol resistance.[39] But the tendency continued to limit the frequency of tribute embassies and to check carefully on their credentials, and there were, as well, the first signs of efforts to limit the size of embassies. The most striking case of these tendencies, and the one that probably did the most to reinforce them, because it seemed to be part of an effective strategy for dealing with the piracy problem, was the reopening of tribute relations with Japan in 1404. A tally system was established; each embassy was limited to 2 ships and 200 men; and there was to be only one embassy every ten years.[40]

There was no real contradiction between this continued tendency to restrict contact and the dramatic expansionist episodes of the Yung-lo period. Those were the products of very special political cirumstances. The Yung-lo Emperor was a usurper, badly in need of the extra scraps of legitimation provided by the exotic embassies to his court. I also suspect that the foreign expeditions were convenient ways to keep actively employed outside China elements that might have menaced the precarious stability of the Ming state if they had been at home and unemployed—sailors, surrendered Mongols, frontier military forces of all kinds. Wang Gungwu has suggested a more positive connection between late Hung-wu restrictiveness toward maritime contacts and the Cheng Ho expeditions; the Hung-wu Emperor had welcomed Southeast Asian embassies as an alternative to illegal trade, but not enough of them had come; the Cheng Ho expeditions were, among other things, a continuation by more active means of this effort to expand legal channels of trade and encourage more embassies to come.[41]

There was one naval expedition into the Indian Ocean after the Yung-lo reign, but beyond that the expansionism of that reign vanished with hardly a trace; in 1473 a literatus opponent of expensive expansionist policies even destroyed most of the official records of the Cheng Ho expeditions.[42] Efforts to establish and enforce rules limiting the size and frequency of tribute embassies continued by fits and starts down to the 1520s, and were especially notable in three periods of great scholar-official influence, early Cheng-t'ung (from 1435), early Ch'eng-hua (from 1465), and early Chia-ching (after 1522). The limitation of Siamese embassies to one every three years may have come before 1435; the same limitation was applied to Champa and

Java in 1436 and 1443 respectively.[43] These were the easy cases. Beginning in 1435, the court also sought to restrict the numbers of people coming to Peking with the annual embassies from the Jurchen and the Mongol Three Commanderies, but the Liaotung officials thought it too dangerous to try to enforce such an order. Around 1440, the court also sought to restrict the size of Oyirad embassies, but the restrictions soon broke down and the embassies became larger than ever; over 2,000 people accompanied the embassy of 1448-1449 to the capital. Around 1465, there were renewed serious efforts to limit the size of embassies from Hami and other Inner Asian trading states, the Oyirads and other Mongol groups, and the Jurchen. Many of those accompanying these embassies did have to remain at the border, but the numbers coming to the capital still were large; in 1496, for example, a Mongol embassy arrived at the border with 3,000 men, and 1,000 of them were allowed to go on to the capital. In 1481, the various Tibetan tributaries were limited to an embassy every three years and to 100 people per embassy.[44] Enforcement of all these limits seems to have slackened after 1488, and especially in the Cheng-te reign (1505-1521), when eunuchs encouraged all kinds of illicit contact with foreign merchants in order to obtain imported luxuries for the Emperor. The succession of the Chia-ching Emperor in 1521 was a major political upheaval, bringing to power an Emperor and scholar-officials determined to eliminate the abuses of the previous reign; here, as often in the early and middle Ch'ing, it seems clear that the domestic political shift was the major determinant of a shift of foreign policy. There was some condemnation of lax handling of Jurchen embassies in 1525; the old quotas were reenforced for some Tibetan rulers; but the tightening-up was directed especially at those tribute relations that were primarily commerical in substance, that is, those with maritime states and with Muslim Central Asia. From 1523 on, Turfan, Samarkand, and other Central Asian tributaries were given tribute intervals of three or five years, and of each embassy 35 at most were supposed to go on to the capital. These limits would have confined to border markets large numbers of Central Asian merchants who had customarily accompanied those embassies to the rich markets of the capital. It would seem that the limits on size were at least occasionally enforced, provoking epidemics of fraudulent documentation; in 1533, it was reported that over 100 individuals had arrived in Peking, all claiming to be kings of Turfan, Samarkand, and even Mecca. Three years later the Governor of Kansu reported that he had 150 Muslim sham kings waiting at the border. But it is not clear that much was done to stop this charade, and Matteo Ricci reported that it was still prevalent around 1600.[45] On the coast, the Chia-ching succession was followed by the rejection of the Tomé Pires embassy because of Portuguese marauding in the Canton Delta and their conquest of the loyal tributary state of Malacca; by the nearly complete rupture of relations with Japan following a brawl between two rival Japanese embassies in Ningpo; and by several decades of renewed efforts to limit maritime foreign trade to that linked to tribute embassies.[46]

In Ming texts on foreign relations, fragments can be found of a coherent rationale for this defensive, passive, and bureaucratic mode of conduct of foreign relations. Foreigners will be attracted to pay homage to China by the obvious superiority of Chinese culture; the Son of Heaven need not, should not urge foreigners to submit, much less force them to do so. China needs a few imports, especially horses from the steppes, but foreigners need much more from China, so that control of foreign trade can be used to make foreigners submissive or at least nonaggressive toward China. China should recognize and confirm the rights of legitimate lines of succession in tributary states, rejecting the claims of usurpers and conquerors. This suits not only its position as moral example and arbiter but also its interest in preserving a multiplicity of powers on its frontiers rather than condoning the rise of one at the expense of the others.[47] The superior sophistication and cleverness of the Chinese also will serve them well in manipulating foreign powers so that they will check and balance each other and do China's bidding.[48]

This is not an inherently implausible set of principles for Chinese foreign policy. China's prestige, its wealth of luxury goods and necessities for export, the interpersonal skills of its elite, even its sluggish but potentially immense military power, were very considerable assets that could be exploited through such a policy. Its bureaucratic government enabled it to maintain a coherent policy and to apply it uniformly at every border point where a particular group of foreigners might seek contact with the empire. But I have not found a single case in which the Ming government turned these assets into anything that could be called a coherent or effective foreign policy. Morris Rossabi believes China was saved from a really serious Mongol menace in the late fifteenth century only by disunity among the Mongols.[49] In the mid-sixteenth century, the Ming mounted no effective diplomatic or military response to Altan's encroachments, and, only after thirty years of debate, military expense, and Mongol raids right down to the wall of Peking, did they work out an arrangement that met the Mongols' rather modest demands for increased trade with China without jeopardizing China's defenses.[50] The platoons of sham Muslim kings were far from the only cases of flouting of the apparently detailed regulations for the control of embassies; in addition to the Mongol cases noted above, the Jurchen seem to have been particularly adept at extorting more presents out of the court by sending large embassies and threatening violence in the streets of Peking, in blatant reversal of the Chinese idea of using Chinese wealth to keep the barbarians docile.[51] The pleas of the Lê kings of Vietnam for Chinese assistance against the Mac usurpers in the north produced, after thirteen years of confused and intermittent debate, a military expedition that obtained a few territorial concessions from Mac Dang-dung; then, when he presented himself at the border as a criminal awaiting execution, the Ming court forgot about its obligation to maintain legitimate lines of succession and accepted him as a tributary.[52] Perhaps the most telling instance of the failure of the Ming government to turn its principles and assets into effective foreign policy was its handling of relations with Hami and Turfan from the 1480s to the 1520s.[53] When the King of Turfan conquered Hami in 1488, the Ming instigated a regional Mongol leader to drive out the invaders. When Turfan troops again took Hami in 1493, a Chinese expedition drove them away. When Turfan occupied Hami for a third time in 1514, the Ming cut off Turfan's tribute. But, when the Turfan ruler attacked

Su-chou on the Kansu frontier, the Ming backed down, acknowledged its de facto control of Hami, and again allowed it to send tribute embassies, so that, by the 1530s, Turfan was among the Central Asian states in whose name many embassies every year were coming to Peking.

How, then, do we account for this yawning gulf between potential and performance in the Ming tribute system? First, the years when its passivity and bureaucratic regulation were most fully developed were those from 1425 to about 1530. This is the least well-studied century in the history of Ming-Ch'ing China, but nothing of what we know of it so far suggests that it was a great age of high bureaucratic morale and energetic statecraft in *any* area of administration. But I think the tribute system had built into it some fundamental obstacles to the making of coherent foreign policy. First, it expressed and reinforced a culturalism that made the study of foreigners and their ways a marginal and not entirely respectable activity. Second, to the degree that its restrictions worked, it limited contact on the basis of which real knowledge of foreign areas could be built up. Cheng Ho and the other leaders of expeditions in the Yung-lo period had accumulated a great deal of good information, but they had few successors. Apart from Mongols gradually assimilating themselves to Chinese culture and temporary visitors on tribute embassies, there were few foreigners in China from whom Chinese could learn about their homelands. Ch'ang-an in the T'ang period had been a marvelous place to meet and learn about all the peoples of Asia; Ming Peking was not. Third, and perhaps most important, to the degree that the tribute system became a system under the Ming, it was focused on ceremonies. As we have seen, concern with ceremonial aspects of the reception of foreign envoys had been an important part of Chinese diplomacy since the Han era, though perhaps no more important than in many other traditional societies. A tendency to focus on ceremonial appearances rather than on the realities of power can be observed in the Sung. Ceremonies are, after all, formalizations of *appearances;* the Confucian hoped that they would not be empty appearances, but he could not imagine social harmony and emotional propriety without their expression in the proper appearances. When a set of institutions had as its primary goal the preservation of a set of appearances, as did the Ming tribute system, it might mobilize resources to bend reality to fit those appearances, as in the Yung-lo period; but, if realities proved inconveniently resistant or Chinese actions embarrassingly incompetent, the logic of the system demanded that appearances be preserved by acquiescing in the realities, as with the Mac of Vietnam and with Turfan's control of Hami, with whatever face-saving arrangements could be worked out, such as Mac Dang-dung's presenting himself at the border as a criminal. Small sham embassies preserved the illusion that the Chinese could regulate foreign visitors, and any kind of embassy was preferable to permittting non-tributary trade in the capital and thus tacitly acknowledging that trade was the sole motive for most of the foreigners who came to Peking. As a last resort, foreign relations could be arranged to minimize foreign contact of any kind with the capital, so that the ceremonial proprieties that would have had to be maintained there would be neither reaffirmed nor infringed.

These strategies of isolation from the capital and to the frontier were especially important in dealing with dangerous foreign challenges in the last century of the Ming. As a result, in these years the tribute embassy became less and less the focus of a comprehensive system for the management of relations with all foreigners. When a settlement finally was reached with Altan in 1570, the regulated exchange of horses for Chinese goods was called tribute but was to take place at the frontier, probably because the Chinese did not want to allow large numbers of Mongols to come to Peking.[54] Japanese tribute relations had been in almost complete disarray since 1523; a final embassy in 1547 was kept under very strict control, just at the time when Chu Wan's efforts to break up illegal maritime trade were leading directly to the great pirate raids of the 1550s. Thereafter, no Japanese were welcome on the China coast; after 1567, limited legalization of foreign trade by Chinese ships provided a new channel, with far fewer political complications, for Sino-Japanese trade.[55] Voyages by Chinese to the new Spanish settlement at Manila made unnecessary what would have been an interesting confrontation of the tribute system with Castilian hidalgo pride and the world empire of Philip II.[56] The Portuguese, declared unacceptable as tributaries after their conquest of Malacca and their early depredations in the Canton Delta, were allowed to occupy Macao from 1557 on, but it seems likely that, for forty years or so, the court thought this was a settlement of more acceptable Southeast Asian peoples.[57] Thus, not only were possible conflicts over tribute ceremonies in Peking avoided and facilities for foreign trade provided at the greatest possible distance from the capital, but the illusion was maintained that the tribute system actually worked and excluded the recalcitrant. Tribute embassies probably continued to be important for a number of relations that were less full of novelty or danger, such as those with Korea, Ryukyu, Siam, and Tibet, but it is hard to tell; there are very few new precedents, very little information of any kind, on any aspect of the tribute system after 1550.[58] This in itself suggests the extent to which the tribute system had lost its vitality and centrality in the late Ming management of foreign relations. The Jurchen continued to send embassies, and to extort favors in connection with them, until after 1600, but the real focus of Sino-Jurchen relations was on their relations with Chinese frontier authorities and on their privileges in border markets, not on anything that happened in Peking.[59] We have no information on changes in the general ceremonial and other routines of embassies in this last century of the Ming, but it seems to me altogether possible that financial stringency, the routine of embassies of no particular political importance year after year, and the prevalence of sham embassies and of blatant commercial motivation for embassies may all have contributed to further reductions and simplifications in the embassy routine. The energetic and realistic statesmanship that produced the single-whip reforms, Chang Chü-cheng's land-registration efforts, and the salt-monopoly reforms of the early 1600s sometimes could be seen in foreign affairs, but largely in those branches not linked to the tribute system, and with few echoes of tributary rhetoric in the policy discussions of the statesmen.[60]

This gap in our documentation on the tribute system and very probably in its energetic management makes it hard to trace in any detail the continuity between the fully developed and comprehensive Ming system of about 1500 and the early Ch'ing system we shall be examining in the rest of this book. But a few broad continuities can be pointed out. The Ch'ing followed the tendency of Han, T'ang, and Ming to assert their sovereignty over Inner Asian border peoples and to assimilate them into semi-autonomous units of Chinese military and administrative hierarchies, and did so with far greater vigor and attention to the substance of power and control than the Ming ever had. They maintained the preoccupation with the formal superiority of the Son of Heaven over all other sovereigns, in ceremonial and documentary forms, that had been especially strong since the Sung and was very much a part of Neo-Confucian political culture. This was a preoccupation with appearances and especially with appearances in the capital; in pursuing their more vigorous Inner Asian diplomacy, the Ch'ing occasionally could be more flexible in these matters when out beyond the Wall. They also inherited, in the Ming archives and printed regulations, the singular idea that all aspects of foreign relations could be and should be managed by the elaboration and application of bureaucratic rules and precedents. This idea affected Ming and Ch'ing management of border markets, dependent border minorities, maritime trade, and so on, even when these were not tightly linked to the institution of the tribute embassy. But the regulations for the management of tribute embassies were among the most elaborately developed, and linked this tendency to bureaucratic management to the concern with ceremonial and documentary superiority in a relation that would form part of the core of the Chinese diplomatic tradition long after the embassy institution itself had become peripheral to many aspects of Chinese foreign relations. Finally, I have suggested elsewhere that, as alien rulers, the Ch'ing hardly could swallow whole the Sinocentric ideology of the tribute system.[61] But it also is intriguing to recall that the rulers of the Ch'ing were descendants of those Jurchen who had been among the most active participants in and skillful and cynical manipulators of the Ming tribute system, a heritage that may have made it seem natural to them at first, but probably also contributed a good deal to their realistic and skeptical approach to it as a tool for the management of foreign relations.

THE EMBASSY ROUTINE

The routine of bureaucratic management and of ceremonies that was applied to our four embassies can be very substantially documented from Ch'ing regulations and from the European sources on the embassies. It has to be seen whole; this is what made the tribute system a system. But, in Chapters 2 through 5, it is only one of many themes. If we get a synthetic picture of this routine first, we can keep it in mind in the case studies that follow without cluttering them with repetitious references to it.

This description is limited to embassy regulations that were enforced for at least one of our four embassies, with additional comments on a few that may have been but for which we have no firm evidence. For each item in the routine, I have added in parentheses the initials of the embassy for which we have firm evidence of it, as, "VH" for Van Hoorn, "S" for Saldanha, "PdeF" for Pereira de Faria, "P" for Paats. I have also added "ChR" for "Ch'ing regulations" where I have a Chinese source for the point but do not refer to it explicitly in my description. The best sets of step-by-step regulations are in *Hui-tien shih-li* and *Ta-Ch'ing t'ung-li;* also useful but less full are *Li-pu tse-li* and the K'ang-hsi edition of the *Ta-Ch'ing hui-tien.* I have noted no significant disagreement among these texts on embassy routines.[62] European documentation is cited in Chapters 2 through 5.

When an embassy arrived in Chinese coastal waters and announced its arrival to the officials in a provincial capital, their first move was to consult their regulations as to the intervals at which embassies were supposed to come from that particular ruler, the number of ships and men that were supposed to bring an embassy, and the authorized port of arrival and route to the capital. If the embassy seemed to be violating any of these rules, the ambassador would be questioned about the reasons for these violations, and his replies would be reported to Peking, but the embassy would not be sent away. It was not really extraordinary for the court to allow an embassy to come to the capital even if it was not at its authorized port of entry or was long overdue. (VH, P, ChR)[63] Before the officials reported the arrival of the embassy to Peking, they generally would insist on receiving for transmission to Peking a copy of the tribute memorial; provincial officials always insisted on controlling foreigners' communication with the capital, and the court would want to have advance knowledge of and the provincial officials' opinions on anything that might be requested in the memorial. The officials also would insist on seeing the ambassador and his presents in the provincial capital. Peking would not approve the embassy until it received the copy of the memorial and the list of presents. The provincial officials insisted on seeing the presents so that they could be sure that they matched the list, and usually saw the ambassador several times, but in the case of Saldanha it was enough for subordinate officials to see him. (VH, S, P, ChR) If they had particular reason to doubt that an embassy really was sent by the ruler in whose name it came, the officials might insist on examining closely the ambassador's credentials or the seal of the tribute memorial. (S) European rulers, however, were not granted Chinese seals to use on their documents as some more regular tributaries were.[64] Officials would be especially wary about all these reporting requirements when, as in the case of the Saldanha embassy, the treatment of the embassy had implications for domestic political tensions and changes, but, even when these sensitivities were less acute, they did not want to risk a reprimand from Peking for inaccurate reporting.

Once the officials had sent off the necessary reports, they relaxed and set out to cultivate with the ambassador the cordial personal relations so important in the Chinese style of politics. They invited the embassy suite to theatricals and banquets, exchanged physicians with them, showed a personal interest in the ambassador and his family. No official was supposed to receive presents from anyone, but the provincial officials' responses to offers of presents ranged from open acceptance through various subterfuges to total refusal.[65] Under the early Ch'ing rules, no trade could be permitted until approval of the embassy was received from Peking. In 1684, it was decided that Siamese embassies would be allowed to begin trade as soon as they arrived; when the Paats embassy arrived, it was not allowed to trade until approval was received from Peking, and it is impossible to tell if this approval was under the old rules or was an extension to the Dutch of the Siamese precedent.[66] The Van Hoorn embassy opened negotiations on trade before it was approved, but without any public proclamation of permission. The negotiations dragged on until after approval was received; probably exchange of goods would have had to wait for the permission even if an agreement had been reached earlier. The trade of the Saldanha embassy was so mingled with Macao's trade and is so badly documented that it is impossible to tell how these rules were enforced or broken in connection with it.

The statutory maximum transmission time for a message from Foochow to Peking was 27 days, from Canton to Peking 32.[67] Embassies rarely rated urgent handling. Even with expeditious handling in the Peking bureaucracy, an answer to a report of the arrival of an embassy hardly could be expected in less than two months. When the approval of an embassy finally was received, there was an abrupt change in atmosphere. Permission to trade was publicly proclaimed, and officials were appointed to supervise the trade. The officials began to send regular supplies of food and money for the sustenance of the embassy party. Most important, the officials began to urge the ambassador to see to the proper packing of the presents and other supplies, to select the people who would accompany him to Peking, and in general to prepare for an early departure. (VH, S) The court might take note and ask questions, and the emperor might even express his impatience, if their departure was delayed. As in their initial reporting on the embassy, and indeed in most of their governing, the officials wished above all to avoid too much or too detailed scrutiny by the court, or even the hint of a rebuke by the emperor himself. One justification for delay is mentioned in the records of both the Van Hoorn and Saldanha embassies; an embassy from a southern maritime country might delay its departure from its port of arrival until early in the year, so that its boats would not be delayed by ice on the northern Grand Canal and so that the embassy suite would not have to buy furs to keep warm. The officials' impatience to get the embassy under way might create new irritations and take some of the bloom off the friendly feelings that had developed in the previous weeks, but it also gave the ambassador a little more leverage in negotiating conditions of trade and any other issues that might arise; if the ambassador suggested that he could not leave until a certain problem was resolved, the officials would make every effort to resolve it, perhaps taking a few risks in the process. (VH)

The Ming authorities had proclaimed limits on the size of embassies over and over, but rarely had enforced them strictly. The early Ch'ing rulers, to judge by our case studies, made them stick. Van Hoorn's suite going to the capital was limited to 24, Saldanha's to 26, and even slaves had to be included in that number. The rest of the party had to remain in the provincial capital, supported by rations supplied by the provincial government, until the ambassador returned from Peking. They could continue to trade there, but goods purchased could not be sent away; the ships that brought an embassy had to remain on hand until the ambassador returned from the capital. (VH, S) I have found no direct statutory evidence for this requirement, much less an explanation of it. Perhaps the authorities wanted to insure that they would not be left having to support a stranded embassy party when its ships departed and others failed to arrive to take the party home. In 1685, the Siamese were granted, as a new concession, the right to send away the ships that had brought an ambassador and send more ships the next year to take him home; it probably was on the basis of this new precedent that the Paats embassy was allowed to do so as well.

The trip from Canton or Foochow to Peking was almost entirely by water, broken only by the pass from Fukien or Kwangtung into the Yangtze watershed. Officials appointed to accompany the embassy were responsible for obtaining from local governments along the way coolies to pull the boats upstream or along the Grand Canal or to carry people and goods over a pass, and provisions for the embassy party. There must have been elaborate regulations and documentation to make this possible, but neither foreign sources nor Chinese compendia give us more than a glimpse of them. The work of the accompanying officials was not always readily accomplished, but still such a water trip, with the embassy party traveling, sleeping, and carrying its goods on a few boats, must have been considerably easier to handle than the problems of coolies, pack animals, and lodgings for an embassy coming by a land route. All expenses were borne by the Ch'ing government; the Dutch were told once that, if they wanted to take goods to Peking to sell, they would have to pay for their transportation, but nothing seems to have come of it. (VH)

Reception by local officials along the way varied greatly according to their individual inclinations and the general political climate. Sometimes officials would refuse presents when the embassy was on its way to Peking but would accept them as it came back (VH), taking care not to precede the emperor in accepting presents which they were not supposed to accept in any case. (ChR) The ambassador himself traveled in the lead boat, which bore a yellow banner proclaiming in Chinese that this was a tribute embassy from such and such a country, and possibly also a flag of his own country or king. (S) Crowds might gather to get a gimpse of the strange foreigners or to watch them pass through the streets on their infrequent and carefully controlled visits in cities.

The trip could take as long as six months; the Paats embassy, traveling day and night, made its return trip in less than two. Arriving at T'ung-chou near Peking, the embassy was met by officials of the Board of Ceremonies and formed a small procession to be escorted through vast curious crowds to their lodgings. These were usually in the Residence for Tributary Envoys (Hui-t'ung-kuan), and were likely to be dilapidated, cramped, and uncomfortable, without sufficient room to prepare the presents for presentation to the emperor. (VH)

The embassy's lodgings were guarded by Ch'ing soldiers, and no one was allowed in or out without official authorization. There was an elaborate set of rules on daily rations to be provided the ambassador and each of his subordinates; the lists of these for two of our embassies are compared in Appendix D. It seems that there was a list of levels of generosity of rations that would be applied to each embassy, but at least in dealing with novel situations like the Portuguese and Dutch embassies the officials had some latitude in deciding how many of the suite would receive the most generous, how many the next, and so on. Chinese regulations devote considerable attention to the responsibilities of bureaucratic agencies for providing for the needs of the embassy, guarding its quarters, and caring for various categories of presents it might bring. If any member of the embassy party became ill, doctors of the Imperial Medical Department were to be summoned to treat him.

The first step in the capital routine was the presentation at the Board of Ceremonies of the tribute gifts and the tribute memorial (always, to the Europeans, of course, just gifts and a letter). Usually this was done on the day after arrival in Peking; Van Hoorn was taken to the Board for these ceremonies even before he was taken to his lodgings. The presents were inspected and some questions might be asked about them and their origins, but it does not seem that they were formally displayed or treated with any special ceremony; the focus of this occasion was on the tribute memorial. A president or vice-president of the Board officiated, in the main hall of the Board offices. The ambassador entered holding the memorial, followed by his suite. All knelt, and the officer in charge of the Residence for Tribute Envoys took the memorial from the ambassador and handed it to the Board officer in charge, who placed it on an altar-like central table. The ambassador and his suite performed the kowtow, then withdrew. (VH, ChR) At this time and in the next few days, the ambassador might be questioned for hours on end by Board officials, especially if, like those studied here, he came from a country of which very little was known. (VH, S) Unfortunately we know only a few of the questions they asked; how far away the country was, what its neighboring countries were, what its products were, what kinds of animals it had. These sound a bit like a few of the categories a gazetteer-compiler would want to cover on a locality.

In the fully developed Ch'ing regimen described in later regulations, the tribute presents had no further function. In our four embassies, this was not quite true. The Van Hoorn presents were displayed outside the Wu Gate two days before the audience.

This probably was one of several instances in which the Ch'ing court still was using, at least occasionally, pieces of Ming ceremony that ultimately were dropped. Also, if there were presents that were especially interesting to the emperor, an appropriate setting would be found for him to view them, as was done for the animals brought by Van Hoorn and by Pereira de Faria. These events seem to have been simple expressions of imperial favor and/or curiosity, without any place in the ceremonial regimen.

The day before the day scheduled for the audience, the ambassador and his suite went to the Board of Ceremonies to practice the ceremonies they would perform. The Van Hoorn records describe this as "going to zombaien before His Majesty's seal." (*Zombaien,* from *ts'an-pai* for the full kowtow, is a good sample of the Dutch penchant for turning Asian words into Dutch verb forms.) They say this ceremony took place behind the main hall of the Board of Ceremonies, and the seal was kept in "a little eight-sided house in the shape of a tower." I can find nothing in Chinese linking any imperial seal to such a location, but oddly enough that tower may be still standing in the Chung-shan Park which occupies the former precincts of the Altars of Earth and Grain. It is eight-sided, is called the Pavilion for Practicing Ceremonies (Hsi-li-t'ing), and one source says it formerly stood in the offices of the Board of Ceremonies.[68] The embassy party went through the whole kowtow twice, to the commands they would hear at the audience, and then went back to their lodgings.

The audience ceremony itself has been described in the opening pages of this book. For all four of our embassies, this ceremony took place in connection with an ordinary audience, which was held every ten days. It would be useful for other students of Ch'ing foreign relations to keep track of the dates of audiences for tribute envoys and to see how many of them, if any, were not on ordinary audience days. (This would apply only to audiences in Peking; the audience and the whole system at Jehol have yet to be studied.) If, as I have suggested, the Ming embassy regimen had gradually dropped some of the spectacular elements that would have done the most to reinforce the mystique of imperial power, this core remained. Only at this ceremony were foreigners on view before the entire higher metropolitan bureaucracy. And, to the extent that the tribute audience conveyed anything to the assembled officials, the very perfunctoriness of the ceremony (in contrast, say, to those for the bestowal of important imperial proclamations, *chao*) and the ordinariness of the occasion must have taught them that the homage of foreigners to the Son of Heaven was natural and expected but scarcely a central prop of imperial legitimacy.

The kowtow in the great courtyard was a highly uniform ceremony; only in the designation of the rank-marker at which the envoy would perform his kowtow was there any possibility of differentiation, and the cases studied here offer very little evidence on this point. When the ambassador entered the T'ai-ho Hall and was given a cup of tea, it was considered a sign of favor if the emperor stopped and spoke to him as he left the Hall. But the more interesting signs of imperial favor were a variety of occasions after the formal audience. Saldanha had two interviews at the Ch'ien-ch'ing Gate; Pereira de Faria was given a banquet in the imperial presence, invited to visit the imperial pleasure parks (probably not in the imperial presence), and the "tribute memorial" he brought was read aloud to the Emperor in Chinese, Manchu, and even in Portuguese; Paats had at least one interview with the Emperor, who heard a translation of some kind of memorial he had presented, asked questions, and listened to European music. The banquet in the imperial presence was a possibility in the Ming regime, but in the Ch'ing was regularly given only to Mongol vassals.[69] The hearing of the tribute memorial may have been a survival of Ming practice that was not finally incorporated into Ch'ing routine. The rest seem to have been personal gestures, without any standing in the bureaucratic routine. Saldanha's interviews are the best documented and seem to me to be the most likely to have been intended to communicate political favor for an embassy and its associates. Paats's are the least well documented and the most puzzling. Pereira de Faria's invitation to the pleasure parks probably was intended as a sign of favor; similar favors were granted to several eighteenth-century embassies, and they may have become commonplace.[70]

At some time after the audience the emperor ordered the bestowal of gifts on the embassy party and the ruler who had sent it and of three banquets on the embassy party. The banquets were held at the offices of the Board of Ceremonies, probably in the main hall. An incense table was set up on the north side of the entrance; as they entered, the ambassador and his suite faced north and performed a full kowtow to thank the emperor for his favor. The first banquet was given in the name of the emperor, the other two in the name of the Board. The only distinctions were that the banquet in the emperor's name began with each person receiving and drinking on his knees a cup of imperial wine and that the officials wore their court robes.[71] A president of the Board presided, seated at the east end of the hall, farthest from the entrance. The ambassador and his suite were seated on his right, other officials on his left. This maintained the placement of the envoys on the less-honored military side, as in the audience, but it also placed the embassy suite facing *south*. This was, of course, the most honorable orientation in Chinese directional symbolism, and in some contexts an indication of sovereignty. It is startling to find foreigners placed in such a position so close to the palace, but Chinese regulations explicitly confirm it, and the chart accompanying them matches perfectly the Dutch description of Van Hoorn's banquets and the plate in Dapper derived from a drawing done on the spot. Clearly the ties between directional symbolism and state ceremony were not simple.[72] According to the regulations of the Imperial Banqueting Court, banquets for tribute envoys were served according to the allotments for the lowest level of Manchu court banquet.[73] The accounts of the Van Hoorn and Saldanha embassies mention some good fruits and Chinese dishes, but are especially striking in their depiction of the dirty silver on which they were served, the big joints of meat, and the primitive table manners of the high officials. At the end of the banquet, the embassy party again faced north and performed a full kowtow to give thanks to the Emperor.

The next important ceremony also was a demonstration of imperial bounty, the presentation of the gifts to the ruler who had sent the embassy, the ambassador, and his suite. Lists of these gifts take up a great deal of space in the compilations of regulations for embassies, and it would be useful if someone would go through more of this material and see if there are any patterns in it. The available information on our four embassies is tabulated in Appendix D; it shows certain obvious regularities, such as the gift of 300 taels of silver to the tribute ruler, and much variation which may be random in the quantities of silks given. It is probable that the larger quantities given the embassy of Bento Pereira de Faria are an indication of imperial favor to it. These gifts were bestowed in the court outside the Wu Gate. A table was set up on the east side of the central path, and the gifts were displayed on it. The Ambassador and his party were seated on a carpet facing north. They performed a full kowtow at the beginning of the ceremony. Then Board officials explained the presents to them, and brought them, first bringing the presents for the King to the Ambassador, then the presents for the Ambassador and on down through the suite. Each person received the gifts kneeling, and then the ceremony closed with all performing another full kowtow. (VH, ChR)

After its audience each tribute embassy was permitted three to five days of trade in its lodgings. The bureaucratic compendia contained minute prescriptions for the regulation of this trade so that there would be no illicit exports nor any devious practices that might lead to disputes between Chinese and foreigners. This was a very important part of the bureaucratic and political heritage of the Ch'ing tribute system, but European ambassadors generally were not very much interested in it. Of our four embassies, it seems that only Van Hoorn's traded in Peking, and it did so only because Peking officials would not accept presents and it thus had goods to dispose of. Its records do show that Board officials controlled the trade very closely and allowed only cash transactions.

For the Europeans, negotiation, not trade, was the goal of an embassy. All came hoping to discuss their affairs with responsible officials and come to agreements embodied in some kind of imperially approved document. But, as far as we can tell, only Paats had any kind of face-to-face discussion with responsible officials, and that only on what was for him a side issue. Paats also managed several exchanges of documents, repeating points or adding new requests as some issues were decided; Van Hoorn was kept completely in the dark about the response to his requests while he was in Peking and even when he left. Saldanha, on Jesuit advice, did not try to turn in a detailed memorandum concerning Macao, but at one interview the Emperor made it clear that he already had been well informed about its affairs by the Jesuits. Pereira de Faria apparently did turn in a detailed set of requests, but left Peking before any decision had been reached on them. In addition to the restriction imposed on embassy movement and contact with officials by long traditions of defensiveness and fear of spying and disorder in the capital, a major hindrance to negotiation was the fact that tribute ambassadors were temporary visitors, and had no time to build up relations with individuals or to learn how the political system worked. A further hindrance was the Ch'ing officials' striking lack of interest in explaining to foreigners anything about their government, their practices in foreign relations, or even the decisions they just had taken about a particular group of foreigners.

The last act of an embassy in the capital was to go to the Board of Ceremonies to receive the imperial edict to the tributary ruler and the communications from the Board recording the gifts sent to the ruler and decisions on other matters. This was a peculiarly perfunctory ceremony; the ambassador faced north, received the documents kneeling, and probably performed a kowtow. Van Hoorn received his documents sealed and could not even learn the substance of the decisions recorded in them; Paats was allowed to receive the documents unsealed. But there was no possibility of reopening discussions once these documents had been received, and the embassy had to leave the capital promptly; Van Hoorn was hurried out of town the day after he received the documents. The return journey to the point of entry was managed in much the same way as the journey to Peking. In Foochow, the returning Dutch ambassadors were treated very coolly; there were good reasons for this in the case of Van Hoorn, but I see none for Paats. The contrast to the cordial reception of an arriving embassy is striking and puzzling.

Much of the detail in this study is drawn from European sources, and we shall have a good deal to say about European planning for tribute embassies and attitudes toward them. It is surprisingly difficult, however, to get any picture of Chinese views of the system that goes very far beyond the bureaucratic compendia. We shall note some poems and short prose passages about the four embassies. They fall into three categories. First, some repeat pieces of the bureaucratic precedents on the embassies. Second, some quote all or part of the Chinese translation of the memorial sent by a tributary ruler, apparently finding it admirable in style or sentiment or both. Third, an essayist or poet may comment on one of the presents brought to the emperor, especially a fine sword or an exotic animal. But one has to look hard for such passages—many of them were found by the indefatigable Fu Lo-shu—and I think no one would claim that the tribute embassy is an important theme in Chinese literature.

Nor is it in the literature of Chinese statecraft and political thought. I suspect that foreign relations in general were a small and declining theme in late imperial political thought, in striking contrast to early modern Europe. I see no reliable way, in the absence of a subject index, to measure the amount of material on foreign relations in the Ming collections of statecraft essays; in the most famous Ch'ing collection, there is no attention to the tribute embassy as such and only about 6 chüan out of 120 on any aspect of foreign relations.[74] In Ming and Ch'ing collections of regulations and precedents concerning ceremonies, the regulations for tribute embassies never take up more than 16 percent of the total, and that percentage is inflated by long lists of presents brought by and given to various embassies.[75] Chinese statecraft could be hard-headed about realities, intensely dogmatic about ceremonial proprieties. But, in late Ming and Ch'ing times, neither attitude led to much sustained attention to the tribute embassy.

Thus, it should be no surprise that, if we depended completely on Chinese sources for knowledge of these four embassies, we would be done with them in thirty pages at most. It is the European sources that make this full study possible. In them, the idea of subordination of other sovereigns to the Son of Heaven is only one of many worries. The Europeans found the restrictions of the tribute system real enough, but the idea that the subordination and the restrictions both were part of a system with a tradition behind it never occurred to them. What they saw was a long succession of difficulties, restrictions, and hindrances to communication that began as soon as they reached their port and culminated in the frustration of negotiation at Peking. Chinese sources tell us almost nothing about the handling of these embassies away from Peking or about the issues about which the foreigners hoped to negotiate. European sources make it clear that we cannot just focus on the embassy in Peking, that what happened in Foochow or Canton was as important in the politics and negotiation of the embassy as what happend in Peking. The ceremonial side was more focused on Peking, but here too the reception of the embassy in its port, its passage through the country, and its welcome by officials along the way were parts of a great ceremonial pageant.

In addition to paying attention to each embassy from its arrival in its port to its departure thence, we will see that we can only understand the goals of the Europeans and the politics of their reception if we set them in a variety of contexts; the evanescent Dutch-Ch'ing alliance of the 1660s, the bare survival of Macao in the face of coastal evacuation and persecution of Catholicism, the opening of non-tributary trade in the 1680s. Our understanding of these contexts, in turn, depends very heavily on European sources. We find a great deal of evidence in these sources that the Ch'ing officials were thoroughly aware of these political contexts and were responding to them in their management of the embassies. We lack Chinese sources on these matters, not because the officials were not interested in them, but because they had no place in the formal bureaucratic categories of the handling of an embassy. Thus, in the Chinese records as well as in the ceremonies, appearances of Chinese suzerainty and continuity of institutions were preserved. The Europeans, sending embassies in part because of their own illusions about what they could accomplish, reinforced at this crucial period the survival into the eighteenth and nineteenth centuries of Ch'ing illusions about the relevance of the tribute system and the view of relations among peoples it embodied to the management of relations into the West.

chapter two

Pieter van Hoorn

1666-1668

The Van Hoorn embassy is the best-documented of our four, the best non-Chinese source of the detail on embassy procedures drawn together in Chapter 1. The systematic records, the thorough planning and generous financing, the presents assembled from all over Europe and Asia, the thoughtfulness and intelligence of the Ambassador and his second in command, all are reminders that the Dutch East India Company was one of the organizational wonders of the seventeenth century. But, by the time the Batavia authorities sent this embassy, they had spent four years and much blood and treasure seeking the friendship of the Ch'ing rulers and secure trade in China, and had gotten little but frustration. I have explained elsewhere the reasons for their disappointments.[1] But it is important to see from the outset that, despite their very substantial investment in this embassy, the people who sent it did not have very high hopes for it. And they were right.

The tribute system did not always function systematically. In the events described in this book, it did so most clearly for the Dutch in 1666; if they had not sent an embassy in that year, it seems probable that they would not have been allowed to trade at all. But Ch'ing bureaucratic management of the embassy made new difficulties for the trade, and the report to Peking of the arrival of the embassy at Foochow probably led to a fairly thorough review of the records of relations with the Dutch, records indicating a good many reasons why they should be viewed as dangerous and unmanageable and excluded as far as possible. The ensuing decision to revoke privileges previously granted to the Dutch might not have been taken if there had not been shifts in court politics, but the review that led to it might not have taken place at all if the embassy had not come. The officials in Foochow also got themselves into considerable difficulties as a result of their handling of the embassy, at least partly as a result of their own failure to keep the Dutch informed about the rules of the system—a general failing of Ch'ing bureaucrats dealing with foreigners which was especially striking and especially serious in its consequences in this case. Their difficulties and the revocation of the trading privileges, quite independent in origin, probably both contributed to the very reserved and restrictive treatment of the embassy on its way to Peking and in the capital. We must keep in focus everything from high-court politics to some long-drawn-out quarrels and confusions in Foochow if we are to understand the court's responses and non-responses to this embassy. But the nature of the Dutch requests and the skill of the Dutch envoys had no appreciable effect on the outcome.

SHUN-CHIH BEGINNINGS AND REGENCY CONFLICTS

In explaining any foreign-policy decision we must give some attention to internal politics. How much weight we can give to internal factors varies with the political system being studied. For a small state with intense foreign relations and broad elite awareness of diplomatic issues, domestic power factors are a necessary part of the causal complex behind any diplomatic decision but hardly ever can be shown to have been a sufficient cause. For a system like that of late imperial China, where very few members of the elite thought much about diplomacy or had much experience of it, we often will find the primary explanation, even the sufficient cause, for a foreign-policy decision in domestic politics. This already has been suggested in relation to the Yung-lo period and the revival of tributary restrictions in the 1520s. But it is surprising how often this holds true *in detail* in the events discussed in this book. Often we shall see changes that are in some measure oriented to the actualities of foreign relations but whose *timing* can be explained only when we look at shifts in domestic politics. Shifts in court politics are important–the Dorgon Regency, the confusion and partial Sinification of the 1650s, the Oboi Regency, the personal rule of K'ang-hsi–but so are the main stages in Ch'ing relations with the maritime power of the Cheng family.

Ch'ing administration of the tribute system along their maritime frontier began in December 1646, when armies under Prince Bolo entered Foochow following the collapse of the regime of the loyalist Lung-wu Emperor. They found there tribute envoys from Ryukyu, Annam, and the Spanish at Manila, all carrying documents directed to a Ming court. Probably all had been sent to the Lung-wu court at the instigation of Cheng Chih-lung, leading supporter (and then betrayer) of that court. All three embassies were sent on to Peking, given presents and sent to carry edicts informing their rulers that they would have to turn in their old Ming credentials and make a proper submission to the new dynasty, and then would be accepted as tributaries. The Spanish apparently never sent an embassy. Ryukyu sent tribute in 1649, Siam in 1652, Annam not until 1661, after the collapse of Ming loyalism in Yunnan.[2]

With the Ch'ing conquest of Canton in 1647 and re-conquest in 1650, the Portuguese presence at Macao was sanctioned by the court without the dispatch of an embassy.[3] Thus, the first important and well-documented contact of Europeans with the Ch'ing tribute system was not Portuguese but Dutch. The Dutch had been important actors in the maritime Chinese world since the 1620s,[4] but had come to the attention of the Ming court only as the frightening "red-haired barbarians" who burned villages and attacked shipping on the Fukien coast in 1622-1624 and had to be expelled from the Pescadores in 1624. Thereafter, their trading center was on Taiwan, outside Chinese jurisdiction, and Chinese traders went there to trade with them. Thus, the Dutch, like the Spanish at Manila, were insulated from diplomatic contact with China by the growth of maritime trade in Chinese hands. Dutch Taiwan prospered greatly as the Japanese withdrew themselves from maritime trade and then expelled the Portuguese from Japan, leaving the Dutch and the Chinese as sole carriers of Sino-Japanese trade and Taiwan as one of the leading entrepots for trade among China, Japan, and Southeast

Asia. But, by 1650, the Dutch on Taiwan were finding their trade with China hampered by the warfare of the Ming-Ch'ing transition, and the Gentlemen Seventeen, the ruling council in the Netherlands of the Dutch East India Company, were urging their subordinate Governor-General and Council at Batavia to consider an embassy to the new Ch'ing rulers in Peking.[5] Thus, when Martin Martini, S.J., arrived in Batavia in 1651 or 1652 on his way to Europe (where he would have his great atlas published by Johan Blaeu in Amsterdam) and told the Dutch that the "Viceroy in Canton" had opened trade to the Portuguese and had said he would "grant a free and unrestricted entry" to any other foreigners who wanted to come to trade,[6] the Batavia authorities were sufficiently interested to order the Council on Taiwan to send an exploratory voyage. Accordingly, a Dutch party under Frederick Schedel was in Canton from 29 August 1652 to 19 March 1653. They encountered both the opposition of "Chinese philosophers," who argued that the Dutch always had been hated and never allowed to trade, and that of the Macao authorities, who repeated the Chinese references to the Dutch attacks of the 1620s, paid three years' back rent on Macao, tried to bribe the officials, and tried to subvert the negotiations through an interpreter called "Emmanuel de Lucifierro." (He was probably of mixed ancestry or a Chinese convert to Catholicism.) Despite all this, the Dutch received written permission to trade from the Feudatory Princes Shang K'o-hsi and Keng Chi-mao, carried on a reasonably profitable trade, much of it with the Princes themselves, and were told they could pick out a site for a permanent trading post. But when Schedel took his leave of the Princes, Keng told him that a high official who had come from Peking advised against granting the Dutch permanent trading rights without consulting the Emperor, and the Princes wrote to the Taiwan authorities urging that an embassy be sent to Peking. When the Taiwan authorities ignored this and sent another exploratory voyage in the summer of 1653, it was peremptorily turned away. It is not clear from Dutch and Chinese sources what the court's response to reports of their 1652 visit had been; probably the Canton authorities had been told that they were very dangerous, and were not to be allowed to trade, but, if they sent an embassy, that was to be reported to the capital.[7]

The striking change from the mid-1652 reception to the mid-1653 rejection probably was a result partly of a general revival of central-government coherence and ability to govern after the post-Dorgon turmoil and partly of a growing tendency to apply the precedents and attitudes of Ming management of foreign relations. Here the key change may have been the result of court debates on the reception of the Dalai Lama, who was on a great progress through Mongolia, apparently at the invitation of the Ch'ing court. The Emperor, with the support of his Manchu officials, wanted to go out beyond the frontier to meet the Dalai Lama, not wanting to support his entourage of 3,000 during a visit to Peking and fearing that if he was not properly greeted a breakdown in Ch'ing negotiations with the Khalkha Mongols would result. But the Chinese officials argued that it was not appropriate for the Son of Heaven to go out to greet a lesser ruler, and on 5 October 1652 the Emperor canceled the proposed journey.[8]

The Batavia authorities did send an embassy in 1655, under Pieter de Goyer and Jacob de Keyser, and, on the recommendation of the Feudatory Princes, the Dutch were accepted as tributaries.[9] The Canton Princes were eager to trade with the Dutch, and apparently expected to profit greatly from Dutch payments for their favorable reports to the court and from taking a cut on Dutch presents which they would transmit to court officials. The frank involvement of high officials in trade with foreigners, either through direct investment or by favoring certain merchants in return for a share of their profits, was characteristic of all Ch'ing relations with the Dutch and the Portuguese down to the mid-1680s,[10] but there was little in the K'ang-hsi period of the frank soliciting and acceptance of bribes we see in the 1650s. By the time the Dutch Ambassadors left for Peking in March 1656 they had promised the Princes in writing that, if they obtained "free trade" for the Dutch, they would receive 35,000 taels. In Peking the Dutch proposed that they send an embassy every five years and be allowed to trade every year in a coastal port. The Manchu officials seemed ready to agree to this, but the Chinese officials reportedly wanted a nine-year interval between embassies and trade in a port only in conjunction with an embassy. The idea that they might be allowed to trade in years when no embassy came soon faded, and the discussion centered on the interval between the embassies. A memorial from the Board of Ceremonies proposed a five-year interval; the imperial rescript in reply changed it to eight. Johann Adam Schall von Bell, S.J., using his considerable influence with the Emperor to protect Macao, life line of the missions, from Dutch competition, took credit for this change and probably did have something to do with it. But, again, it is important to notice a broader policy context. In 1652-1654, the Ch'ing court had negotiated for the peaceful surrender of Cheng Ch'eng-kung, but these negotiations had broken down, and, beginning early in 1655, there were signs not only of increased Ch'ing military activity against his forces but also of effort to shut off the trade between Ch'ing and Cheng areas that was so vital to the Cheng regime. On 6 August 1656, a few days after the Dutch embassy reached Peking, an edict forbade all maritime trade in Chinese ships.[11] The Ch'ing authorities must have known that Cheng traded with the Dutch on Taiwan, and thus would have seen anything more than minimal tribute-connected trade with the Dutch as an unacceptable leak in their economic blockade of the Cheng regime.

The Dutch embassy had reached Canton in July 1655, dickered with the Princes and officials and awaited approval from Peking, headed north in March 1656, reached Peking in July, had their audience delayed until October 2 by the death of an imperial prince, performed all the usual tribute ceremonies but found very few opportunities for negotiation, reached Canton in January 1657 and left in February, nineteen months after they arrived. They had spent about f100,000 (over 28,000 taels) on presents and on the journey north. And, after all this time, effort, and expense, they took away with them a reply that meant that in the eyes of the court they had no business on the China coast until 1664.

But Canton was far away, and the Feudatory Princes still were in power and in business. In 1657 and 1658, the Dutch carried on a small amount of trade with client-merchants of the Princes at The Bogue. In 1659, Dutch merchants stopped briefly in the same area but then decided to take their cargoes on to Taiwan; in 1660, the Dutch were absorbed in plans to reinforce Taiwan and possibly attack Macao; and, in 1661, the situation of both the Dutch and the Ch'ing was transformed. The Shun-chih Emperor died on February 5, and was succeeded by the boy K'ang-hsi Emperor, with the government in the hands of a regency of four Manchu officials. Cheng Ch'eng-kung invaded Taiwan on April 30, and Casteel Zeelandia finally surrendered on 1 February 1662.[12] From 1662 on, the application of tribute rules to the Dutch became part of the complex story of Ch'ing-Dutch negotiation and naval cooperation against the Cheng regime.[13] When the Dutch sent a fleet in 1662, hoping to cooperate with the Ch'ing in attacks on their common enemy, the court gave permission only for them to sell the cargo they had brought on that fleet. When a larger fleet came in the fall of 1663, the Boards of Ceremonies, War, and Finance recommended that the Dutch again be allowed to sell the cargo they had brought, but that the proposals of the Fukien officials that they be granted some kind of permanent permission to trade in years when no embassy was sent be rejected. But later, probably after news of the great victories of November 1663 reached Peking, they were given permission to trade every other year. In March or April 1664, very much in the style of the tribute system, rewards of silk and silver were sent to the commander of the Dutch fleet, and in July it was decided that similar rewards should be sent to the Governor-General in Batavia; but, although court envoys came all the way to Foochow to ascertain the correct name and titles of the Governor-General, apparently this decision never was carried out, probably because the Dutch failed to send an embassy in 1664.

It is hard to make sense of these fits and starts in rewarding the Dutch except by assuming that the court was sharply divided on policies toward them, and that friendship and reward carried the day only in the wake of the impressive Dutch contributions to the war against the Cheng regime. The citing of Ming precedents by Chinese scholars probably carried less weight under the "Manchuizing" regency than it had at the Shun-chih court, but the reduction of this negative factor was more than balanced by harsher coastal blockade policies, culminating in the evacuation and devastation of much of the south China coast, and by fearful reactions to the size and firepower of the Dutch ships. It is interesting, however, that the Siamese also got permission to trade without sending a tribute embassy in one or two years in this period.[14] It is possible that the concessions to them and to the Dutch owed something to the efforts of Fukien and Kwangtung officials toward finding alternate channels of maritime trade to compensate for the strict prohibition of trade by Chinese ships.

The eight-year period after the first Dutch embassy ended in 1664; that was the year in which the Dutch were supposed to send a second. They did not, and this marked the first step in a process in which failure to comply with the rules of the tribute system became a source of irritation, although probably not the most important one, in the waning Sino-Dutch entente. They did not send an embassy primarily because they did not see themselves as participants in a tribute *system* centered on embassies at regular intervals, and the Ch'ing rulers apparently had done little or nothing to explain this concept to them. Nor had the rulers, in their letters early in 1664, reminded the Batavia authorities of the obligation to send an embassy. They did remind the Dutch in Foochow in conversations, but the Batavia authorities took no note of the records of these oral reminders. Possibly the Foochow rulers were hedging their bets a little, not putting anything in writing that would seem to be promoting a regular long-term relation, preferring a strategy of using and then excluding this alarming naval power.

The Ch'ing court probably had assumed that the Dutch would send an embassy in 1664, would trade in connection with it, and that the newly granted biennial trade privilege would take effect two years after that. In the fall of 1664, the officials at Foochow did not seem very much upset about the failure to send an embassy, and the court gave permission for the Dutch to trade "in order to buy provisions" for the planned Ch'ing-Dutch naval expedition against Taiwan. Once again the officials reminded the Dutch orally that an embassy must come the next year, but said nothing in their letters to Batavia. But, when no embassy came in 1665, they seemed very much upset, saying they had written to the court that the Dutch would send an embassy and would be embarrassed by the failure to do so. They did recommend to the court that they be allowed to trade under the biennial privilege; the court approved this, but also ordered that the Foochow authorities find out why no embassy had been sent. Another source of Ch'ing annoyance with the Dutch in 1665 was related to the bureaucratic-managing mentality of the tribute system if not to the embassy institution itself. Each tributary had its designated port of entry, and was supposed to limit its contact to that port, and generally to one coordinated arrival at that port per year or per embassy. But, in 1665, the Dutch sent ships to Foochow at two separate times, also sent a ship to the Chang-chou estuary, and had ships arriving at Foochow from cruising expeditions on the Chekiang-Fukien coast and from their outpost at Keelung. The ship near Chang-chou and the later arrivals at Foochow were not allowed to trade; the other arrivals led to no end of bureaucratic difficulties for the Dutch and for the aging and ill Governor-General Li Shuai-t'ai who had to cope with them. The officials were particularly uneasy about the Dutch re-occupation of Keelung at the north end of Taiwan, fearing that it might be a step toward an alliance with Cheng Ching, based in southern Taiwan. The Dutch raid on P'u-t'o-shan and the evaporation of any possibility that there might be another expedition to Taiwan in which Dutch cooperation would be desirable were other causes of the breakdown of Sino-Dutch relations that year. Ch'ing recognition of the breakdown was articulated within the tribute system; the court sent orders to officials all along the coast that no more Dutch ships were to be allowed to put in to Chinese ports until an embassy was sent and that, if any did so, the people on them would be punished, perhaps by death. The biennial trading permission was not revoked; it was granted to the Siamese in 1665,[15] and probably still had a place in the efforts of the Fukien and Kwangtung rulers to keep some channels of maritime trade open. Li Shuai-t'ai died early in 1666, leaving a deathbed memorial expressing his fear that the Dutch, with their incessant coming and going, would come back to make more trouble. But apparently the court responded to this by ordering the Fukien officials to give the matter careful consideration and wait another year to see if the Dutch were upright people and to see if an embassy came.[16]

DIFFICULTIES IN FOOCHOW

On 12 January 1666, after letters from Foochow had made it clear that further relations with China would be impossible without an embassy, the Governor-General and Council in Batavia decided to send one. Later they had second thoughts, pointing to the great expense and meager hope of satisfactory results. But the Gentlemen Seventeen still expressed high hopes for trade with China, even in a letter sent in November 1665 and received in June 1666. The Governor-General and Council agreed that a less closely restricted trade with China would be a fine thing if it could be obtained and that one final attempt had to be made. They still deplored the great expense, but hoped that much of it would be covered by the profits of trade conducted in connection with the embassy.[17]

The Governor-General and Council decided that it was essential to the prestige of the embassy that one of their own number go as ambassador, and Pieter van Hoorn volunteered for the position. This interesting man, by far the most competent and thoughtful of our four ambassadors, was the son of a regent of the city of Amsterdam. He and his older brother had been partners in the trade in and manufacture of gunpowder. The brother died in 1649, just after the Treaties of Munster and Westphalia ended the Thirty Years' War and thus very probably reduced the international demand for gunpowder. In 1652, Van Hoorn made an excellent marriage that linked him to one of the most influential and cosmopolitan family networks of the great city. His wife, Sara Bessels, was the granddaughter of Gerrit Reynst, who had planned a China voyage in 1598 and had been one of the founders of the Dutch East India Company and its second Governor-General, and niece of Gerrit Reynst the Younger, owner of an opulent house on the Keizersgracht full of Indian and Roman curiosities. Among her cousins were the Witsens, Baltic grain-traders and experts on Muscovy. In 1663, Pieter van Hoorn sold his gunpowder mill to the city for demolition, and took office with the East India Company.[18] His excellent family connections probably were the key to his immediate appointment as a member of the Council of the Indies in Batavia.

Curiosity and love of adventure probably moved him to volunteer for a task that an ordinary old Indies hand would have shunned as devoid of opportunities for private profit and overfull of decorum and dickering with devious heathens. Van Hoorn requested and was granted permission for his thirteen-year-old son Joan to accompany him. Joan later rose rapidly in the service of the Company, and was Governor-General from 1704 to 1709.[19] Constantijn Nobel, who had been involved in Sino-Dutch relations since he was Secretary of the Council on Taiwan in the late 1650s and was the Company's most astute and politic negotiator with the Chinese, was to be Van Hoorn's first counselor, and was to complete the embassy if Van Hoorn died. Justus Six and David Harthouwer, both of whom had served in Foochow in previous years, were to be left in charge of trade there. The embassy suite was to include, at Van Hoorn's request, a junior merchant, a secretary, a surgeon, two assistants, a cook, an artist (the first sign of Van Hoorn's interest in systematic collection of information), six bodyguards, six soldiers, and two trumpeters with silk flags on their trumpets.

The best summary of the goals of the embassy was a list, in Van Hoorn's instructions, of points to be raised in negotiations in Peking:

1. He was to remind the Ch'ing officials of the acceptance of the De Goyer–De Keyser embassy, and that since then the Dutch had become much better known to high and low, and had been generally seen to be men of honor and good faith, who "have no other goal in China than to earn a living honestly through commerce; trade being considered honorable by all the world, since it makes all kingdoms, lands, and states to flourish and prosper."

2. This embassy is sent to congratulate the present Emperor

upon his accession, to ask that the Dutch be accepted as friends of the empire as they were in his father's reign, "and that now the desired trade in China may be granted and allowed to us once and for all, with more freedom and assurance than heretofore, perpetually and without limitation of time."

3. The Peking authorities should note that there was no good reason not to grant these requests, since the trade would be to the benefit of China as well as of the Dutch; of these benefits only a few small samples had been seen so far.

4. They must complain that, so far, their trade in China has not been very profitable, since it had been hindered by the rulers of the maritime provinces, who sometimes forbade trade "on the pretext that it was forbidden by the court in Peking," so that sometimes whole shiploads of goods had to be sent away unsold, and, even when trade was allowed, it was restricted in many ways; they could not go on trading in this way, and asked that new orders be sent for the conduct of this trade.

5. The Dutch performed great services for the Ch'ing against the Cheng regime, and the Fukien officials had promised them compensation for these efforts, but finally had given them nothing, saying that nothing could be accomplished without the sending of an embassy.

6. Now that they had brought this embassy from so far away with such costly presents, they expected that, in satisfaction of the earlier promises for compensation for their efforts, they would be granted free trade.

7. They estimated the expenses of their three years of war fleets at about 600,000 taels, for which they never had the least compensation.[20]

The last three points reflected the conviction of at least some of the Dutch at this time that they had been promised but had not received reimbursement for the expenses of the 1662-1664 expeditions. Actually, of course, these expeditions had been sent in pursuit of Dutch goals, especially revenge on their Cheng enemies and the possible reconquest of Taiwan, as well as to aid the Ch'ing, and I see no evidence that in 1662-1664 the Dutch expected, asked for, or were promised any such reimbursement. They now were prepared to drop their demand for it if they got "free trade," but would revive it in the peculiar circumstances of a Ch'ing embassy to Batavia in 1680.[21] There is no evidence that Van Hoorn ever mentioned the claim for reimbursement in China.

In Van Hoorn's instructions, the nature of the "free trade" the Dutch sought was spelled out, and some of the restrictions practiced by the provincial rulers alluded to, in three points which the Dutch would like to see in an "imperial letter":

1. Permanent free trade throughout the Chinese empire and along all its coasts, with permission to trade in all its major ports, including Canton, "Chinchieuw" (the Chang-chou area of Fukien), Foochow, Ningpo, "Nanking" (the lower Yangtze ports), and others, bringing ships every year, establishing permanent factories, importing and exporting all goods without exception.

2. Permission to buy from and sell to anyone, without being forced to trade with particular merchants. (The most persistent Dutch complaint of the early 1660s was that they were allowed to trade only with the client-merchants of the high officials.)

3. That all provincial officials be strictly forbidden to hinder their trade in any way, so that the Dutch may enjoy the benefits of the Emperor's concessions.

In projecting such comprehensive concessions, the Governor-General and Council probably saw no reason to settle in advance for anything less, and probably also had in mind their experience with the Mughal empire of India, where they did trade fairly freely in various ports and inland trading centers and where their privileges were confirmed by imperial decree; in these instructions they even used the Mughal term *firman* to refer to the kind of imperial confirmation of trading privileges they hoped to obtain in China. It is not clear what their priorities were among these demands if they could not be met in full. Although they had been refused trade in the Chang-chou area in 1665, from 1662 to 1665 they had been allowed to trade in Foochow, even though Canton was supposed to be their port of entry for tribute. If the court had been disposed to grant any concessions to them, it might have considered allowing them to trade in more than one port; when trade finally was opened to

Europeans in 1684, it was permitted in Canton, Amoy, and Foochow. They objected very seriously to the restriction of their trade to the client-merchants, but effective action against it was unlikely while the Feudatory Princes remained powerful in Fukien and Kwangtung. There is no way of knowing what the Dutch would have considered sufficient concessions to make continued trade with China worthwhile; I think that, if they had been allowed to continue trading at all, they would have done so, complaining and arguing all the way.

Van Hoorn was to argue that permission to trade was simply fair compensation for Dutch naval assistance against the Cheng regime, but he also was authorized to make additional concessions if necessary to gain trading privileges. An annual embassy to Peking, preferably by the chief of a coastal trading post at the end of the trading season, as in Japan, was a possibility. But it would be less expensive to keep a resident agent in Peking and send gifts for him to present to the court every year—a casual and ephemeral anticipation of the "resident minister" issue that was so important in the nineteenth century. He might agree to Dutch abandonment of Keelung, but only after three years of satisfactory trade, and might agree never to ally with the Cheng regime, again on condition of satisfactory trade with the Ch'ing. But if the Ch'ing sought further naval aid against the Cheng regime he was to promise nothing, but simply say that the Dutch might send help at their own convenience. All this was speculative; none of these issues ever seems to have come up in Van Hoorn's negotiations.

The Batavia Councillors had before them as they planned for this embassy the records of the De Goyer-De Keyser embassy and of Ch'ing-Dutch relations in the early 1660s, which contained ample evidence of the applications to the Dutch of the ideas, ceremonies, and bureaucratic regulations of the tribute system. They noticed some of the evidence, but did not perceive the system. They did not even consider Ch'ing decisions that their embassies should come via Kwangtung in no more than three ships; they sent five ships to Fukien. They ordered Van Hoorn to conform to the established ceremonies and usgaes of the court, "for, as we already have experienced enough, opposition accomplishes little with that willful people." They specifically noted that it probably would not be possible to present letters and credentials directly to the Emperor, and that at the audience they might neither see nor hear him, but simply perform an "obeisance in the Chinese manner" before him. The rulers of the Dutch Company had their share of the prickly sense of rank and precedence so characteristic of seventeeth-century Europeans, but, unlike their Portuguese contemporaries, they did not have to defend the personal honor of a sovereign in their dealings with Asian courts; there *was* no ruling prince in the Netherlands, and, in any case, they were representatives of a trading company, not of the state. Dutch diplomats did not worry very much about the calisthenics that accompanied Oriental diplomacy—head to the ground at the Mughal court, face to the tatami and no looking around before the Shōgun in Japan—as long as they did not interfere with the substance of diplomacy.[22] Any remaining worry about the implications of an embassy to China and its ceremonies probably was assuaged by the explanations offered by Constantijn Nobel. In 1665, the Peking authorities had noted that the Dutch were supposed to send an embassy every eight years, and had ordered the Foochow officials to find out why they had not sent one in 1664. Commenting on a translation of this edict done in Foochow by Victorio Riccio, O.P., Nobel simply ignored this clear statement of the requirement of periodic embassies, and said the Dutch probably would be able to trade every year if they sent an embassy to the court every year, as they did in Japan. He noted that Riccio had used "tribute" to translate the Chinese *chin-kung,* "which nevertheless means nothing more than sending an embassy and presents, or coming to pay respects, without any meaning of subjection or vassalage, for the Persians, the Siamese, the Tonkinese, the Mughals, and Muscovites, and still more, all being great princes, also sometimes send their ambassadors to Peking, all of which are included in the word *chin-kung.*"[23] Thus they did not worry about the ambiguities of appearances or the nuances of Chinese views of the embassy as long as they knew that in reality embassies were accepted from independent as well as subordinate rulers; the Ch'ing rulers, in turn, were in fact frequently willing to overlook the realities of power as long as appearances were preserved.

The embassy carried presents valued at about f100,000, or 25,000 taels, largely for the Emperor but including some for the Foochow officials and some for Van Hoorn to give at his discretion to high officials in Peking. The presents for the Emperor included amber, coral, spices, incense woods, Persian carpets, fine sword blades, fine pistols, telescopes, ornamented glass lamps, and copper statues. But the presents that made the greatest impression in Peking, and carried on that long continuity from the "Hounds of Lü" to Cheng Ho's giraffe and beyond, were four Persian horses and two small oxen from Bengal.[24] The letter to the Emperor was brief, wishing him "every blessing in Your glorious succession" and asking that the Dutch be accepted as friends of the empire and allowed to trade freely. More substantive requests and complaints were to be made by Van Hoorn in Peking; this procedure was adopted because it was common in Asian diplomacy to write only a short formal letter to the sovereign and especially because the Emperor was a minor and not in charge of the government, so that all business would have to be done with his Councillors. The Dutch may not have realized that they had hit on the only way they could communicate to Peking their complaints about the Foochow officials' restrictions on trade; the Foochow officials would insist on seeing a copy of the letter to the Emperor so that they could forward it to the court, and if there was anything in it detrimental to them they surely would find some way to stop the embassy.

In Foochow, Van Hoorn was not to have anything to do publicly with Dutch trade, leaving it under Nobel's supervision, but was to keep well informed about its progress and was not to leave for Peking until it was well under way; if trade was not permitted, he was to refuse to go to Peking, and take the trading goods to Keelung or Nagasaki. Once in Peking, he was to conform to the usages of the court, not be in a hurry, and do all he could to procure a favorable decision. The Governor-General and Council even authorized him to spend up to 10,000 taels in bribes in Peking to obtain free trade, and sent along the necessary silver.

The embassy's five ships sailed from Batavia on 3-4 July 1666. They carried, in addition to the presents, merchandise valued at f486,038, the highest total of Dutch imports to China in any one year between 1662 and 1690. The ships arrived off Foochow on August 5. In the previous four years, the Dutch had had much experience and many difficulties with the geography of this area.[25] There was an excellent harbor among the islands in the mouth on Min River which they called "Dutch Harbor." Two of the ships went in there, but some always would stay outside, at Hsiao-ch'eng, where the Ch'ing would find it hard to impede their departure. Beyond "Dutch Harbor," Dutch ships or boats had to pass inspection at the fortress on Min-an-chen and make their way up the Min, past the wide reach later called "Pagoda Anchorage," to Foochow. Contact soon was established with the Foochow officials, and messengers began to move back and forth. The first Dutch emissaries reported that Chang Ch'ao-lin, the new Governor-General, had asked if the Dutch intended to take any action against the Cheng regime. The Dutch said they had no such intention, and there is no evidence that the subject was brought up again.

Nobel and several others had been in Foochow before and knew the Feudatory Prince Keng Chi-mao and Governor Hsu Shih-ch'ang. But none of them really appreciated the change in Ch'ing attitudes as a result of the events of 1665. None knew anything of the bureaucratic difficulties of an embassy, and, because of poor interpreters, vague explanations by the Chinese, and occasional differences among the officials as to what was required, they would continue to find them hard to understand. Nobel went up to Foochow on August 9; Van Hoorn planned to stay on board until he saw how things were going. But soon Nobel reported that nothing could be done until the Ambassador had been received there, and, on August 23, Van Hoorn and his suite went up the river. At Min-an-chen, the officers in charge wanted to inspect their baggage, but backed down when Van Hoorn refused. On the next day, Van Hoorn went to greet Keng Chi-mao, who had music played to welcome him, seated him in the place of honor, and, in addition to a great many expressions of pleasure and esteem, urged him to turn in to Governor-General

Chang a copy of the letter to the Emperor and to make arrangements to have the presents for the Emperor brought up to Foochow as soon as possible. Then they were very hospitably received by Chang Ch'ao-lin, who urged them to bring the presents up soon so that they could be shown to the officials; then, he said, they would write to Peking. On the 25th, Chang sent a pass to allow boats to move up and down the river to fetch the presents. Van Hoorn told the official who brought it that this could not be done until trade was permitted, but the official said it must be done at once, and, if it was not, the officials could not write to Peking.

If Chang had stuck to his insistence that the embassy could not be reported to Peking until the presents were seen, the embassy would have been delayed by over a month. The Dutch could hardly have been less well prepared to show the imperial presents; they were divided among all the ships, some in "Dutch Harbor" and some out at Hsiao-ch'eng. Some had been packed under the trade goods for greater safety. The trade goods that had to be unloaded to get at the presents included large quantities of pepper that had been shipped in bulk and had to be sacked as it was unloaded. Having already had trouble with pilfering when goods were moved on Chinese craft, the Dutch insisted that the presents be brought up on one of their ships and on their own boats, and this led to more delays and discussions with the officials. The moving of the horses and oxen, including their transfer from one ship to another, caused still more difficulties. The horses and oxen were shown to the officials on September 8, and the last of the presents reached Foochow on September 15.

But Chang soon gave way on his demand. On August 26, Van Hoorn gave Chang and Keng copies of the letter to the Emperor and the list of presents for him. Both insisted that the presents and the trading goods must be brought ashore as soon as possible, but discussions of procedures and vessels to be used made it clear that this would be no easy matter. On August 29, Keng and Chang sent a memorial to the court in Peking reporting the arrival of the embassy. We know this only from the citation of this memorial in the court's reply; there is no evidence that they told the Dutch that they had sent such a report. What would they have done if on inspection of the presents they had found that the list of presents they had reported to Peking was not entirely correct? Such difficulties were most likely to arise about the horses and oxen, very intriguing to the court and the Emperor and not replaceable from stocks of trading goods and presents for the officials. I suspect that, by August 29, Keng and Chang had reports from trusted Chinese officers who had seen the horses and oxen; Chinese junk captains had been involved in efforts to move them from Hsiao-ch'eng to "Dutch Harbor" on August 18-19. Although the rest of the presents had reached Foochow by September 15, they were not shown to the officials until October 14. The Dutch were annoyed by the delay, and I find it puzzling. The presents were viewed at the same meeting at which the officials welcomed five imperial envoys who had come from Peking; perhaps the Foochow officials had put off viewing the presents so that these officials could see them doing so and report this to Peking. In the Oboi Regency years, officials frequently were sent from Peking, ostensibly to carry an important document or to investigate a particular matter but, in fact, also to check up on the general conduct of provincial administrators. The presence of such capital officials in and around Foochow probably affected the provincial officials' dealing with the Dutch for the rest of 1666, and certainly aggravated the consequences of the departure of Dutch ships early in 1667. Imperial envoys may have been sent on tours of all southern coastal areas late in 1666; some of them contributed to the miseries of Macao at this time, as we will see in the next chapter. But I have found no Ch'ing documentation on coastal policy or the dispatch of such envoys in these months.

The very cautious handling of the embassy did not prevent a good deal of personal cordiality. What better way to try to control and understand a group of strange and dangerous people than to rely on Chinese skills and human feelings? Van Hoorn, in turn, seems to have been doing all he could to be courteous and conciliatory in order not to jeopardize this last attempt to get satisfactory trade with China. Governor-General Chang

Ch'ao-lin sent his physician to treat Van Hoorn, and Van Hoorn sent his physician to Chang. Keng Chi-mao seemed especially interested in young Joan van Hoorn, introduced some young daughters of his son, and asked if Joan was already married; we are reminded of Ch'i-ying's interest in Sir Henry Pottinger's family in 1843. Keng at first refused all presents, but eventually saw fit to accept some in the form of presents from Joan van Hoorn to his granddaughters. In contrast to Keng's caution, Chang Ch'ao-lin almost never refused a present, and frequently sent emissaries to inquire about the purchase of goods, especially coral; the Dutch usually responded to these broad hints by making him a present of the desired item. The Dutch also had time for sightseeing, visiting a monastery on a nearby mountain. Van Hoorn even found time for a conversation with "Captain Carvalho," a Macao Christian Chinese who was commander of Keng's bodyguard of "black soldiers," probably a remnant of the Cheng bodyguards of escaped Macao slaves. Captain Carvalho told Van Hoorn that the mysteries of the Christian faith could not be properly expressed in Chinese, and that Chinese converts who knew no European language could be judged only on their works of charity and righteousness, not on their knowledge of the faith.[26]

From the beginning, the problems of bringing the presents up and viewing them were linked to those of bringing trade goods up and getting permission to sell them. The Dutch had understood that, if any embassy was sent, trade would be permitted in connection with it; this was true, but of course it depended on the acceptability of the embassy, and officially nothing could be done until approval of the embassy was received from Peking. But discussions of trade could and did proceed. Early in September, client-merchants of the high officials offered to sell the Dutch raw silk at the very high price of 250 taels per picul, and said that, since the export of silk was strictly prohibited, the deal would have to be concluded promptly, before the expected arrival of high officials from Peking. Nobel refused, apparently simply because the price was too high. Keng and Chang waited until all the presents had arrived in Foochow, and then said that negotiations on terms of trade could begin, but no public proclamations of permission to trade with the Dutch could be posted until approval of the embassy was received from Peking. Without these public proclamations, there was no chance at all that any merchant without official connections would try to break in on the client-merchants' monopoly. On September 26 and 27, the Dutch and the client-merchants agreed that trade would be on the riverbank near the great bridge as it had been the year before. Price negotiations began only on October 16, after the high officials had seen all the presents for the Emperor. As usual, they began with wide differences between prices asked and offered and progressed with many fits and starts. A price agreement finally was reached on November 28, after approval of the embassy had been received from Peking.

A number of side issues complicated the process of unloading and selling the Dutch imports. On September 16 and 19, the Dutch had mentioned to Keng and Chang that they hoped to send one or more ships back to Batavia soon. The officials knew that any such departure would be a violation of Ch'ing regulations, but they did not inform the Dutch of this, saying that they must at least wait until the imperial decision on the embassy had been received. Early in November, their worry about this seems to have combined with the presence in the area of the above-mentioned imperial envoys and perhaps with anticipation that other capital officials might arrive with the imperial decision on the Dutch embassy to produce sharp reactions to a number of Dutch irregularities. The Chinese official had been annoyed by visits to the walled city by various Dutchmen trying to carry on a private trade in small quantities of pepper and sandalwood. After November 3, no Dutchmen were allowed within the walls without specific permission in advance, and, at the same time, the Dutch took strict measures to cut off this trade in violation of the Company's monopoly. A small ship arrived from Keelung, and the officials insisted repeatedly that it would have to leave at once. The Dutch wanted to send down river the small yacht *Bleiswijck,* which had brought the presents up to Foochow, but Chang Ch'ao-lin, probably aware that this was the ship they wanted to send to Batavia, found various pretexts for not issuing the necessary pass. On November

3, the *Bleiswijck* set out down the river without a pass, but four Ch'ing war-junks forced it to return to Foochow. The Dutch insisted that their "compradors" (here in the original sense of buyers of provisions) had told them a pass was waiting for it at Min-an-chen, but the Foochow officials refused to allow it to leave.

A more protracted if not necessarily more serious source of difficulty was the Ch'ing involvement in the affairs of some Chinese who had come from Batavia on the Dutch ships. These people, some of whom seem to have been coming home to China to stay after long years at Batavia, had brought along quantities of pepper, sandalwood, and silver that do not seem to have been large but were in excess of the Company's stringent regulations. The Dutch had confiscated these goods when they were discovered, and the Chinese, some of whom were left completely destitute, had appealed to the Chinese officials. Despite the prohibitions of maritime trade and the general official hostility to Chinese who lived abroad, the Foochow officials indicated that these people would be released on security pending a decision from Peking. Chang Ch'ao-lin first tried to persuade the Dutch to give the Chinese their goods, and then on December 23 ordered them to do so, and on December 29 said that no ships might leave until the Chinese goods were brought ashore. The Dutch then said that they would not turn over the goods, but, if Chinese officials would go to the ships, they would show them where they were and would not hinder them from taking them. This was carried out without further incident, but it is not clear if the unfortunate Chinese managed to get their goods from the officials.

On November 11, Governor Hsu Shih-ch'ang sent the first news of a decision in Peking concerning the Dutch. This actually was a copy of a second imperial decision, reached on October 13 upon recommendation of the Board of Ceremonies. The Board pointed out that Kwangtung was supposed to be the route for Dutch embassies, but this one had come to Fukien. The Emperor (that is, the Regents in his name) decided to overlook this irregularity at least for this time. Limitations on the size of the embassy party were recalled, and it was stated that, if the Dutch wished to bring trading goods to Peking, they would have to pay for their transportation, as the Siamese had done recently. On November 18, Chang Ch'ao-lin sent a copy of the more important decision reached on October 11, expressing gratification at the arrival of the embassy and confirming the Dutch privileges of sending an embassy every eight years and trading every two years.

On November 19, the Dutch visited Chang to thank him for writing to Peking on their behalf and to offer him the presents from the Batavia Council. Chang and other officials now seemed willing to accept the presents after the Emperor had received his, and at the same time very anxious that the embassy should leave for Peking as soon as possible. Chang said the embassy party would have to leave Foochow for Peking within ten to twelve days. Van Hoorn said he would prefer to wait until February, to avoid the cold of the northern winter. He also pointed out that Nobel, who was in charge of trade, was to accompany him to Peking, and would not be able to do so until the trade was largely completed. This statement increased the officials' incentives for a quick trade agreement, and the imperial approval of the embassy made the trade legal. At the same time, the Dutch were becoming more eager to reach a trade agreement because they feared that further delays would make it impossible to send off their ships with return cargoes before the end of the north monsoon. The difficulty of victualing the ships and their poor condition because of constant worm damage in Chinese and Southeast Asian waters made the Dutch very reluctant to have their ships remain there until the fall of 1667. And, as far as we can tell, the officials still had not told them that Ch'ing regulations forbade them to send their ships away before the Ambassador was ready to leave. The client-merchants made a new offer on November 21. After hard and complex bargaining, an agreement was reached on November 28, and a placard permitting trade was brought. Prices of Dutch imports were a little lower than the year before, probably because the Dutch had brought such large quantities. Prices of Chinese exports were about the same as in 1665, and raw silk was an exorbitant 200 taels per picul if any of it could be obtained at all. The trade moved fairly smoothly. The pepper in the Dutch lodge was weighed out to the merchants, the *Bleiswijck* was allowed to go down to get more goods, two ships were allowed to come up to the Pagoda Anchorage for more efficient unloading, and some Chinese goods were delivered to the Dutch, who even managed to obtain a small quantity of silk, a prohibited export, but much of it of inferior quality. Much of the trade was with the client-merchants of the high officials, but we cannot be sure that all of it was. On January 8, accounts were settled; the Chinese merchants still owed the Dutch 34,000 taels, the collection of which would occupy the party left behind in Foochow for the whole time the embassy party was away. In summary, the arrival of the embassy had made the trade legal, but otherwise it does not seem that precedents for the management of trade in connection with embassies had much effect; the Dutch still had to extend credit, and the trade was not open to all Chinese merchants.[27]

On December 23, the officials informed the Dutch that boats, horses, coolies, and so on were ready for their journey to Peking, and asked them how soon they could be ready. It is not clear why there was such a delay after Chang's warning that they should be ready to leave about December 1. Perhaps the transportation arrangements had taken longer than they expected. Perhaps they had been waiting to see how the dispute over the goods of the overseas Chinese would be resolved. The Dutch began to prepare for the journey, and the officials began urging them to hurry; on 9 January 1667, Keng told them he already had learned that the Emperor and the Board of Ceremonies were displeased that they had been delayed so long. This is implausible; there had been just barely enough time for another exchange of communications with Peking after the receipt of the edicts approving the embassy. This sudden shift back to impatience for departure may have been linked to a momentous and mystifying shift in the court's policies toward the Dutch. In the same interview on January 9, Keng told them that, when they first arrived, their biennial trading privilege had been confirmed, but shortly thereafter another edict had been received revoking the privilege. The Dutch did not receive, then or later, any hint of an explanation of this reversal. Perhaps the change was in some way a result of the arrest on December 12 of Sunahai, who had been one of the chief shapers of coastal policy in the early 1660s.[28] Keng said he and Chang were going to plead for a reversal of this decision. Certainly a reversal, making possible fairly regular Dutch trade in Foochow, would have been in their interests. They may have thought they had a better chance for reversal if the Dutch reached Peking as soon as possible and made a good impression there, and this may explain in part their haste to get the embassy party under way. (Or they may have simply wanted to be rid of these troublemakers who no longer were in favor at court.) There was no reversal, and, as we will see, the revocation was confirmed after the embassy reached Peking. But the Dutch received no explanation of the confirming edict and mistranslated it; so they never did understand that on 9 January 1667 they already had known of the decision that had ended their chances of regular legal trade in Chinese ports for almost a decade. The door had slammed shut on the Dutch, but the sound had not reached them.

The Ch'ing officials' haste to send the embassy off now collided head-on with Dutch haste to send their ships back to Batavia. On December 30 and 31, after dodging the issue for over three months, Chang informed them that no ship might leave until the Ambassador returned from Peking. The Dutch protested that they had to send at least one ship immediately to carry news of the progress of the embassy, and, on January 6, they thought Keng said that, if they could not wait any longer, they should just send the ship off quietly and not tell anyone. On January 9, he again insisted that no ship would be allowed to leave, but by then the Dutch had acted on their understanding of his statement of the 6th, and the ship had gotten safely out to sea. On January 10, the officials complained that this ship had threatened to fire on Ch'ing junks that had tried to stop it. He said that a high official who had been in the Tinghai area surely would report this departure to Peking, and the Foochow officials would have to answer for it. The Dutch wanted to send off at least three of the remaining four ships, but, on January 12, Chang insisted that no ship might leave until he wrote to Peking and got permission; he said a reply could be obtained within forty days (a very optimistic estimate), but the Dutch insisted that even that would be dangerously close to the end of the monsoon. They repeated this in a letter and added a list of the materials needed to repair their ships, most of which could not be obtained at Foochow. The departure of the embassy party originally had been scheduled for January 14, but had been delayed by the discussions about the ships. On January 18, two officers came to ask why the embassy had not left on the 14th. Van Hoorn replied that they were waiting for a reply to their letter of January 16 about the ships. This must have seemed to confirm official fears that the embassy would not leave until the ships had been allowed to go, although it is not clear that the Dutch intended to make such a threat. Apparently such an answer had been expected, and Keng and Chang had instructed the officials how to reply to it. The officers said their superiors had decided that the ships would be allowed to leave. Subsequently the officials made no effort to hinder the departure of three ships, but insisted that the *Bleiswijck* remain at Foochow, so that it could not sneak away and, presumably, so that at least one ship would be on hand to take the Ambassador away.

Early in May 1667, the Dutch who had been left behind in Foochow heard rumors that Chang Ch'ao-lin and Hsu Shih-ch'ang had been suspended from their offices pending investigation by an imperial commissioner from Peking, and Keng Chi-mao had been fined 2,000 taels for having allowed the Dutch ships to leave without obtaining approval from Peking; an "extract from an imperial letter" translated in the Dutch records placed the primary blame on Chang for having "misled" Keng.[29] But "Captain Carvalho" insisted that this was a pretext, that the punishment of the officials actually was the result of their treatment of a private agent (a eunuch?) of the Empress Hsiao-chuang, the powerful grandmother of K'ang-hsi and a leading opponent of Oboi. According to this story, the agent had been sent to Foochow to check on conditions and trade there, and had orders not to divulge his commission. Chang had suspected that he had such a commission, had tried to persuade him to divulge it, and, when he would not do so, had thrown him in prison. The agent had appealed unsuccessfully to Keng, and had remained in prison for several months, until word was received that he had indeed had such a commission. Chang apparently was dismissed from office. But, even if the departure of the Dutch ships was only a pretext for the dismissal of Chang and the fining of Keng, it is interesting that the Ch'ing authorities thought it was a sufficient pretext. Also, officials at Canton in 1668 knew about it and expressed fears that they would be punished if they allowed Dutch or Portuguese ships to depart from their jurisdiction without authorization from Peking.[30]

TO PEKING AND BACK

The embassy party left Foochow on January 21, after complying with official demand that it be reduced to 24 people, including slaves. We have a full day-by-day record of the embassy's journey to Peking, sometimes fascinating, sometimes tedious, defying summary, providing good information for students of many topics in the history of the period. The party went up the rivers of Fukien and over the Hsien-hsia Pass to Chekiang, then down-river to Hangchow, and thence north along the Grand Canal. On January 28 and 29, at Yen-p'ing, they greeted and gave presents to Chou Ch'üan-pin, their leading opponent in the battles of November 1663, who had surrendered to the Ch'ing in 1664 and was now in command of the garrison there. He advised the Dutch not to trust anyone too much in Peking, and to do all they could to get the Regents on their side. From February 9 to 21, they were held up at P'u-ch'eng-hsien in the far north of Fukien, while the officials tried to round up enough coolies to take their baggage over the pass. They then set out across the pass, observing the many monasteries on the mountains and, on February 26, were once again on boats and on their way down the river in Chekiang. On March 3, they reached Lan-ch'i-hsien, the first place since Yen-p'ing where the local official saw fit to welcome them; certainly an interesting contrast to the many official welcomes of the 1656 embassy. In addition to being affected by the generally closed and sensitive

EMBASSY ROUTES TO PEKING

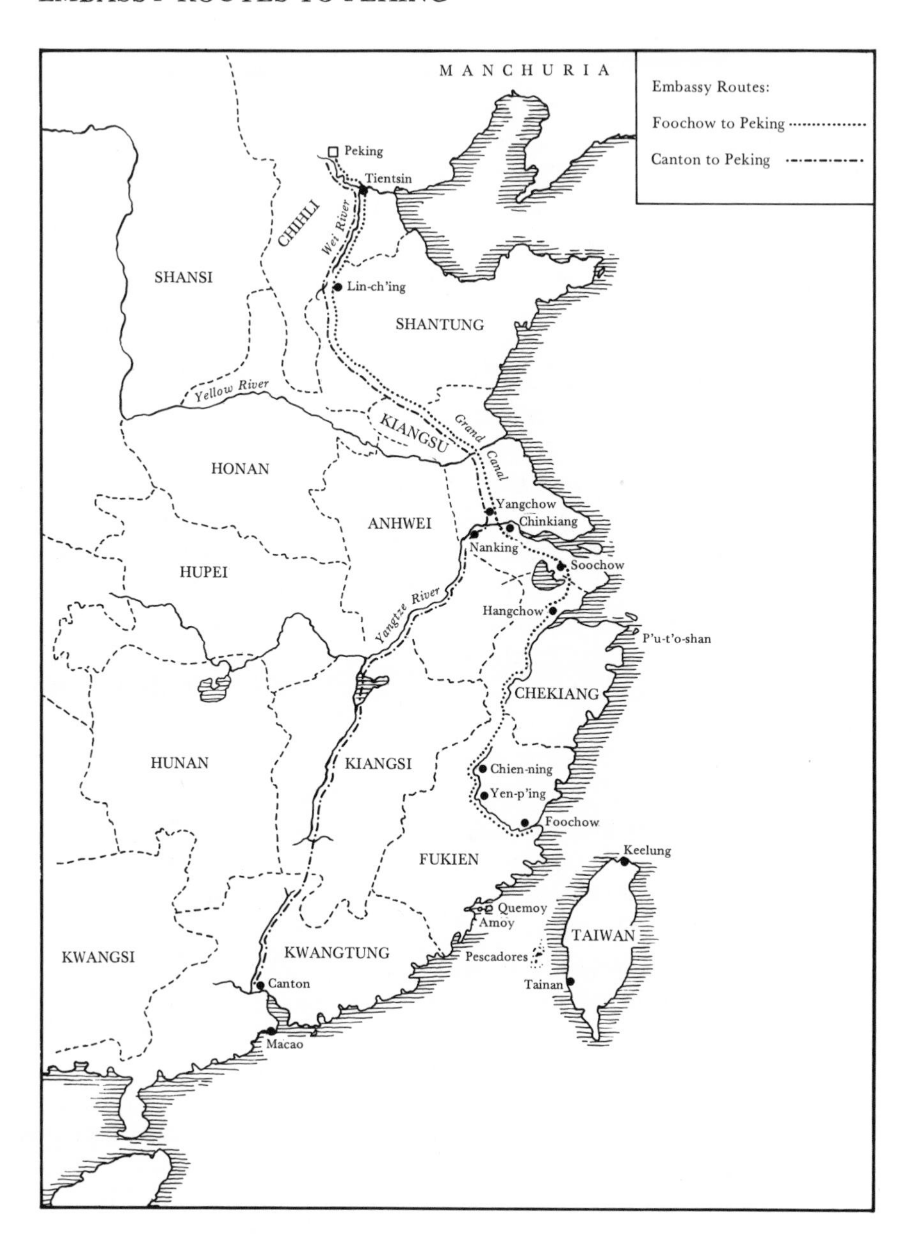

political atmosphere of the time, some of the officials probably already knew that the Dutch trade privileges had been revoked, and they may have known something of the troubles over the ship departures.[31]

On March 9, the embassy reached Hangchow, where they spent ten days in efforts to arrange for boats for the next lap of their journey and had a series of very interesting interviews with the officials. Van Hoorn was anxious to explore the possibilities of Dutch trade in that area, and the results of his commercial investigations were very encouraging; silk was said to be available for 100 taels per picul, and the Dutch verified this by buying a catty for one tael; pepper was said to bring 17 to 20 taels, and sandalwood 40 to 50, or even 70. Explorations of the prospects that the Dutch would be *allowed* to trade there had more equivocal results. The Dutch were cordially received and banqueted on March 13 by the Governor-General, who refused their presents but later hinted that he might accept them as they returned from Peking. Van Hoorn said he hoped that any Dutch ships that might be blown into shore in the Hangchow-Ningpo area would be treated well. The interpreter at first seemed unwilling to translate this, and, when he did, the cryptic reply was that this was the Emperor's land and his commands would be obeyed. But, when the same question was repeated two days later, the Governor-General said that any Dutchmen who found themselves in that area would be treated as friends. Van Hoorn made plans to send one of the 1667 ships to try to trade in Chekiang, but was not able to carry them out.

The embassy party left Hangchow on March 19 and reached Soochow on March 23, where they exchanged presents with the high officials. They left there on March 31, having been delayed again by the officials' inability to find enough boats for the next lap of their journey. Apparently, a new set of boats had to be obtained at specified points, and the previous set sent back to the point from which they had started. On April 6, they reached Chinkiang, where they were invited to a banquet by the Manchu commander of the garrison, who was so fascinated and curious that he forgot to eat. He came to their boats the next day to see the horses and oxen. "The music instruments were also played at his request, which pleased him wonderfully well. He looked them all over carefully, and especially the organ; as if he could not understand where the sound came from."

On April 10, the party crossed the Yangtze to Kua-chou, and entered the Grand Canal. On the 16th, they reached Huai-an, where they waited seven days while a new set of boats was assembled, and were banqueted by the Director-General of Grain Transport. On April 25, they crossed the Yellow River, which flowed south of the Shantung peninsula at this period, and began a much slower and more difficult part of their journey. The Canal rose to cross western Shantung, and there were many flash-locks. A boat coming upstream was winched into the lock over a weir against a down-flow of water by large gangs of coolies; going down, a gate would be opened and the boat would "ride the rapids." The locks filled very slowly, and sometimes long waits were necessary.[32] Sometimes the officials could not find enough coolies to take over when one group had completed its regular route, and progress was very slow indeed. Van Hoorn now saw that the journey would take far longer than he had expected. Arriving at Chining on May 7, they were greeted by several officials and told they would have to change boats again. But, by the 9th, no new boats had been found, and, when the Dutch protested the delay, the officials decided that it was not necessary to change boats there after all. On May 12, they reached Lung-wang-miao, the point where water was channeled into the Canal to flow off to the north and south. Thereafter they were repeatedly held up, once for a day and once for two days, while the gates of a lock were shut to allow enough water to accumulate. Other delays were caused by the crush of big grain-tribute boats, which sometimes could not be passed due to the narrowness and shallowness of the Canal. Arriving at Lin-ch'ing on May 26, they found the last lock into the Wei River closed by iron chains, pending the arrival of a new magistrate, but, by May 31, he had arrived and the party passed into the river, noting wearily that they had passed forty-seven locks in their journey across Shantung and that it had taken them thirty-two days. On June 9, they reached Tientsin, where the commander of the garrison gave them a banquet but refused to accept their presents until they had been received at Peking. The journey up the river to Peking was again slowed by difficulties in obtaining coolies, but, on June 19, the embassy party left their boats near T'ung-chou, and, on the next day, they entered the capital. The journey from Foochow had taken almost exactly five months.

The embassy was greeted not far from the walls of Peking by a junior official of the Board of Ceremonies. They formed a little procession, with a Dutch "Prince's flag" preceding the carts bearing the presents, then the oxen and horses, another flag, two trumpeters, the Ambassador in a sedan-chair, his son on horseback, Nobel, the steward, and the secretary. Inside the gates of the city they stopped and "they thanked God for his grace, that thus far had brought them in good health, on a journey of so many miles, in which six months had been spent, on water and on land [this must include the voyage from Batavia to Foochow], and in which we had passed by or through 37 cities, 335 villages, and 34 pagodas." They were taken directly to the Board of Ceremonies, where they were to turn over their presents and letters. "Everywhere they went the street teemed with people, and thousands lined the sides of the streets, so that it became so dusty (the streets of Peking being much subject to this inconvenience) that they could hardly see out of their eyes, could hardly see the color of their clothes."[33]

At the Board, they were asked various questions, including where their horses and oxen were from and whether Nobel or young Joan van Hoorn was of higher rank. Then they were summoned into another room to turn over to the Presidents of the Board the letter from the Governor-General and Council in Batavia to the Emperor. Van Hoorn managed a number of small victories over Chinese protocol. In the first interview, he refused to sit on the floor, and was given a seat at a table. When he turned over the letter, he removed his hat, placed the letter on a table, and bowed three times, but did not kneel.

The party then was taken to the quarters that had been prepared for them at the Residence for Tributary Envoys. The Ambassador found them hopelessly inadequate and pointed out that there was not even enough room to keep the horses, oxen, and other presents safe and dry. The officials promised that they would obtain better housing on the next day (they finally got it a week later), but said the first order of business was to prepare the horses and oxen so that they might be viewed by the Emperor the next morning. The Dutch protested that this would be impossible, especially since the saddles for the horses and the little cart for the oxen to pull would have to be unpacked and prepared, and their present quarters were too cramped for that kind of work. But, in the middle of the night, more officials came from the Board to say that the horses and oxen must be brought to the palace before daylight, for the Emperor wanted to see them in person that morning; somehow they got them ready.

Board officials arrived about three in the morning to take the party to the palace. I am not sure where they took them; they went through three great gates, out a fourth, along the walls of the "inner palace," then in another gate; that might describe a way up to the Wu Gate, out a side gate and around to the Tung-hua or Hsi-hua Gate and through it. Nobel and the secretary accompanied the animals, and returned to report that they had been inspected first by Oboi, the most powerful of the Regents, later by Ebilun, another Regent, and then the Emperor himself had emerged on horseback from the inner palace to see them. I am fairly sure this is the earliest eyewitness description in a Western language of the young Emperor. He was just beginning to issue edicts on his own, and two months later would formally take charge of the government; but his struggle with Oboi for actual control would not end until June 1669.[34] It is striking that the Dutch, a large and well-fed people for the seventeenth century, thought the Emperor was about sixteen years old, when actually he was only thirteen. They said he was "a young man of middling height, quite white, about sixteen years old, modestly dressed; he wore a blue damask coat with a little embroidery on the front, back, and shoulders, and yellow boots. He looked the horses over especially carefully, and could hardly take his eyes off them, laughing constantly and talking about them to the First Councillor [Oboi]." He asked the Dutch how far Holland was from Batavia and Batavia from Foochow, and who had sent the Ambassador. Even at this early date he had a better grasp of facts than many of his ministers, who sometimes forgot that Holland and Batavia were two different places and who consistently referred to the Governor-General at Batavia as the "King of Holland." The Dutch replied that that Ambassador had been sent by the Governor-General at Batavia in obedience to the command of the Prince in Holland.

Later on the same day, the Ambassador was summoned to attend the inspection of the rest of the presents at the same place in the palace. Oboi and Ebilun came to see them, and asked the Ambassador what kinds of grain grew in Holland and, learning that the horses and oxen came from Persia and Bengal, if the Dutch lived in peace with the rulers of those countries. The presents were packed up again, and the Ambassador left. Two Dutchmen who had remained at the palace to unpack the ornamented glass lamps reported that the Emperor had come to see the lamps and had also had two of his suite sit in the little cart and be pulled about by the dwarf oxen. The Ambassador also was visited by a number of high officials, and, early on the 22nd, a eunuch came, sent by Oboi, to ask if the Dutch had any coral or fine cloth to sell. Van Hoorn replied that the Governor-General in Batavia had expressly forbidden him to sell anything, and thus avoided what might have been a risky involvement in clandestine trade in the capital. Later on the 22nd, this unofficial contact was cut off by the arrival of four officers and twenty soldiers, who guarded the embassy's quarters and refused to let anyone in or out except on official business. The Dutch steward, who was also in charge of the presents, was summoned by the Board of Ceremonies to assist in unpacking more of the presents for the Emperor, and returned to report that a sample of each kind of goods had been chosen to be taken to be inspected by the Emperor.

On the 23rd, the Ambassador and his suite were conducted to the palace before dawn (they had been warned the day before that they should be ready at this hour), to be present at the formal acceptance of their presents. This took place in the great courtyard before the Wu Gate, where on the same day a Ryukyuan embassy was receiving its gifts and edicts. A Board of

Ceremonies official said the Emperor might come to see the gifts. Van Hoorn had been looking for an opportunity to present a brief memorial of thanks directly to the Emperor, and wondered if this might be his chance, but the official said it was not the custom to do so. The Dutch waited for several hours, then were told that the Emperor would not come that day, and that they might leave, but should leave behind their interpreter with the keys to the chests of presents. The interpreter reported later in the day that the chests had been opened and the presents taken in to be viewed by the Emperor.

On June 24, the Dutch were summoned to the Board of Ceremonies to practice the kowtow before their audience. Later the same day, one of the Presidents of the Board of Ceremonies came to the Dutch lodgings, questioned the Ambassador for about an hour, and left behind two subordinates who continued the questioning most of the rest of the day. Unfortunately, only the first few of their questions are recorded in the Dutch records, but apparently they were mostly about Holland, its location, its products, and so on. They even asked, according to the Dutch, what sort of tails the sheep and rabbits had in Holland; they must have been thinking of the fat-tailed sheep of Muslim Inner Asia.

On June 25, the President of the Board of Ceremonies and several of his subordinates arrived at the Dutch quarters shortly after midnight to take them to their audience. The audience has been described in the opening pages of this book; it was purely formal, and the Emperor did not approach them or say a word to them. Later, a Board official came to ask more questions: How far were various places from Holland? Were there tigers and other dangerous animals there?

The Dutch thought that, now that the ceremonies were over, they could get down to business, and, on the 26th, this hope was encouraged when the Board of Ceremonies asked if they had any additional requests to make or any additional presents to give. The Dutch had already drawn up lists of presents they would offer to the Regents, the Grand Secretaries, and the Board officials, and now added a list of "personal" presents from the Ambassador to the Emperor, and a list of their substantial requests. In their requests they decided to omit any complaint about past Dutch difficulties in Foochow and to say nothing more about the departure of the ships early in 1667. They decided to present, in accordance with their instructions, the following requests:

"1. That the Dutch may come to trade in the Ch'ing Empire always and every year, with as many ships as are needed for their goods.

2. That Dutch ships may come to trade at Canton, Changchou, Foochow, Ningpo, and Hangchow.

3. That the Dutch may trade, buying and selling, with everyone, without being forced to sell their goods to anyone against their will.

4. That the Dutch be allowed to buy and export silk and silk goods, and all other goods except those the Emperor is pleased to forbid; of which the Ambassador asks that he may be informed, so that the Dutch may fully obey the Emperor's will.

5. That the Dutch ships may trade as soon as they arrive and freely depart as soon as they are ready.

6. That all provisions and food supplies for the Dutch ships may be bought and taken to the ships.

7. That the Dutch be provided with a secure dwelling in the place where they come to trade, they paying for it, in order to store and sell their goods and keep them safe from fire; for even the *Chinkon* [*chin-kung,* tribute goods] was in danger in Foochow.

8. That the above points be confirmed by sealed letters from the Emperor, so that the (provincial) authorities shall obey them.

If the Emperor agrees to these points, the Ambassador has orders to consult further concerning the *chinkon* [that is, concerning future embassies at regular intervals], at the Emperor's pleasure."

These requests were turned in on June 27, and, from then to the 30th, the Dutch explained, first in writing and then in tedious interviews with Board officials, why they had come via Fukien when Kwangtung had been the route designated for them after their first embassy. The officials seemed satisfied with their replies that they had known nothing of such a requirement, and that they had begun to come to Fukien in order to attack the Cheng forces there. But, soon thereafter, the officials sent orders to Fukien that, in the future, Dutch embassies were to come via Kwangtung; these orders were received in Foochow before August 14,[35] but, as far as I can tell, the embassy in the capital was never informed of them—a truly staggering example of non-communication and the focus on routine bureaucratic categories in the tribute system. The Ambassador's "personal" presents to the Emperor were accepted and taken away, but the Dutch got nowhere with their efforts to give presents to the high officials. They tried again to present written lists of the presents on July 2, but simply aroused more suspicion; they were questioned very closely as to how they had even known of the existence of the Regents to whom they proposed to give presents, and explained that they had seen two of them and had been told that there were two more. They may have aroused suspicion by mentioning the *four* Regents, for, by this time, one of the four was dead of natural causes and another was completely at odds with the dominant Oboi faction and soon would be tried and executed.[36]

The Dutch were visited in their lodgings by several very high-ranking Manchus, including a brother of the late Shun-chih Emperor, another close relative of the Emperor, and "a great lord who has command over the nobility of the Emperor's palace." Several officials came just to ask to hear their musical instruments. The imperial banquets were held on July 12, 16, and 20. At the beginning of each, the Ambassador and his suite kowtowed to show their gratitude to the Emperor. Chinese food was served on dirty silver dishes, and then excellent joints of beef and mutton; the Manchu dignitaries fell to the meat "more like wild than civilized men." Some of the officials had no tables or plates at all, but simply ate from the mats in front of them. One high official asked the Dutch if they had brought any sacks to take home their leftovers. When they replied that that was not their custom, their interpreter brought out bags and cloths "which exuded such a sweet air that we almost became sick from the smell," and took home the leftovers from the Dutch party's dinners.

The Dutch began to fear that they would not learn the imperial decision on their requests until they were ready to leave, and thus that they would have no opportunity for further discussion or clarification. They also were planning to request permission to sell the trading goods that would come on the ships sent to fetch them. They prepared a short note to the Board of Ceremonies asking to receive the Emperor's reply eight or ten days before they left. A Board official would not even receive this document, but insisted on having it translated for him orally, and then said that they would get an answer, either orally or in writing, and also an imperial edict to the Governor-General. This was vague enough, but Van Hoorn simply said that this was what he had sought. When the Dutch sought to present the note at their second banquet on July 16, they were told that would not be necessary, since its contents already were known.

The Dutch had brought along no goods for trade, and had told the officials several times that they had nothing to sell. But now it was becoming clear that the high officials were serious in their refusals of the presents that had been offered. Thus, in order to avoid taking back to Foochow the goods they had brought along for these presents, they decided to inform the Board of Ceremonies that they did have goods to sell after all. The Board first requested a list of the goods to be sold, then sent officials to pick out some of the best of each kind to be bought for the Emperor. The Dutch thought the prices they offered far too low, but said they were willing to agree to any price if it was for the Emperor. Later, the officers returned to say that the Emperor already had better goods of the kinds the Dutch were offering in his storehouses, and would not buy any from them.

Trade was conducted in the Dutch lodgings under the supervision of officials of the Board of Ceremonies from July 28 through August 2. Guards were stationed at the door. Those buying for high officials had to show letters from their masters to gain admittance, and ordinary merchants had to show their silver. When a sale agreement was reached, both parties had to take the goods and the silver to the official in charge and have the silver checked and the agreement approved. No credit sales were allowed. The Dutch were pleased with these conditions, but were not so pleased with the prices. It is hard to compare these prices with either cost prices or prices received in coastal trade; cloth sold is recorded by the piece, and the size of the piece is not always certain; coral and amber varied enormously in cost and value depending on condition and size of pieces; firearms and sword-blades did not usually figure in the coastal trade, but were quite important in this trade. For the cloth, prices seem to have been a little better than in the coastal trade of the 1680s, but not as much as would be expected, allowing for costs of transportation. For some sword-blades and firearms the cost is recorded, and is very little below the sale price. The principal reason for the low prices seems to have been that the first three days of the trade were monopolized by the agents of the Princes and high officials; prices seem to have been somewhat better after trade was opened to ordinary merchants on July 31, but, by then, many of the best goods were gone.

On July 29, the embassy party was summoned to the Palace to receive additional presents for the Governor-General and the Ambassador. Those received on the 17th had been according to the precedents for the Goyer-Keyser embassy; these added 20 pieces of silk to the 60 for the Governor-General, 2 to the 24 for the Ambassador. On the same day, recalling that a Board official had told them that they would receive the Emperor's replies three or four days before they left the capital, they decided to wait until after they had received them to request permission to sell the goods the ships would bring. Then, on August 3-5, their stay came to a dismal and frustrating end. On the 3rd, heavy rains broke down part of the wall of their compound, and a Board official came to tell them they would receive the Emperor's replies the next day and leave Peking the day after that. Pieter van Hoorn had pursued his hopeless task with diligence and tact, seeking new ways to reopen discussions, spending days answering questions about the details of the presents, why they had come via Foochow, what his son's rank was, and so on, reminding the Ch'ing officials that he did not know the protocol of their court, seeking their guidance in conforming to it and, at the same time, fulfilling his expectation of substantial negotiations. Now he gave up, and probed and protested no more.

On August 4, the embassy party received on their knees the imperial edict to the Governor-General at Batavia and three communications from the Board of Ceremonies to the Governor-General. These were sealed, and, when the Ambassador said he would have to open them and have them translated, the Board officials said this was strictly forbidden, and later sent a messenger to inform the Dutch that the Fukien Governor-General had orders to check the seals and to behead the interpreters and writers for the Dutch if they had been broken. They got several different explanations of the contents of the letters, including one that they would be allowed to trade every year, but were unable to get reliable information. On August 5, they left Peking in a dismal little procession, their horses belly-deep in the mud of the streets.

They would not have been much enlightened if they had been able to read the documents. According to Dutch translations of these documents, probably done at Batavia, the imperial edict contained little more than a list of the presents. Of the three communications from the Board, the first simply certified that the Dutch envoys had arrived and had delivered the presents. The second reviewed the investigation of the reasons why the Dutch had come via Fukien, and reported the Emperor's decision that, in the future, they must come only via Kwangtung. The third summarized the Dutch requests, reviewed the regulations for trade in the capital by tributary envoys, and said that if no embassy was sent no trade would be allowed. Then, according to the Dutch translation, it said that, in the tenth month of the fifth year of K'ang-hsi (1666), it was decided that the Dutch might pay tribute every eight years and trade every two years. This date is after the confirmation of the privilege, and about right for the date of the revocation. According to Chinese texts, in the fifth year of K'ang-hsi an edict stated: "Since Holland pays tribute every eight years, its permission to trade every two years should be permanently stopped." (*Ho-lan-kuo chi pa-*

nien i-kung, ch'i erh-nien mao-i yung-yuan t'ing-chih.) Probably the Dutch translation reversed the sense of the passage by overlooking "permanently stopped" at the end of the Chinese sentence. The document as translated then went on to summarize the Dutch requests, permitting them to buy provisions for their ships, and rejecting all their other requests. The mistranslation, juxtaposed with the rejections of Dutch requests to trade, was to leave the Batavia authorities completely puzzled as to whether they had any prospects for trade in 1668. There is no record of any document or conversation in which any Ch'ing official pointed out the revocation or made any attempt to explain it to the Dutch. Mis-translation and non-communication had carried the day.[37]

The return journey from Peking to Foochow took almost a month less than the journey north, largely because there was enough water in the locks of the Shantung section of the Canal. The general at Tientsin and local officials at Lin-ch'ing, Yangchou, Hangchow, and other places accepted presents they had refused when the Dutch were on their way to Peking. The embassy party arrived at Foochow on 2 November 1667.[38]

During the nine months of the embassy party's absence, the small Dutch party remaining in Foochow had devoted most of their time to trying to persuade the client-merchants or the high officials to pay their debts. Chang Ch'ao-lin and/or his clients owed 19,000 taels at one point, but, by the end of August, had reduced their debt to 10,000. In mid-November, the Dutch were still trying to collect about 4,000 taels, apparently mostly from Chang's clients. There is no record of the final settlement of these debts, probably as a result of a copyist's error; if these debts had not been settled, this certainly would have been brought up in later Dutch conversations with the officials and in the embassy's final report to Batavia. The small party in Foochow had also had a great deal of trouble getting passes to take fresh food to the crews of the ships and provisions and food to a ship that was supposed to go on to Keelung. As part of the general tightening of coastal control, some kind of imperial commissioner or inspector was now stationed at Min-an-chen, and was responsible for most of the Dutch difficulties. Then, on

July 29, three ships arrived off Foochow from Batavia, to take the Ambassador home and to sell their cargoes, which were valued at a total of f328,773.[39] These ships were inspected and closely watched by Ch'ing officials, and it was almost impossible to get supplies for them past Min-an-chen. On August 14, the Dutch were told that orders had come from Peking that, in the future, the Dutch were to come every eight years via Kwangtung, and that they were not to be allowed to trade in 1667. The Foochow officials now wanted to have nothing to do with the new arrivals, and, on September 11 and 16, they demanded that all the ships except the one that had stayed over the year before leave the harbor at Hsiao-ch'eng and go off to the north. The ships did leave but did not get very far, and, by about November 1, they were back at Ting-hai, where they were kept under close watch.

When the embassy returned to Foochow, Van Hoorn was received very coolly by the high officials, and was informed that no further trade could be permitted there, not even of the remainders of the goods that had been brought in 1666. The new Governor-General would not even accept his letters attempting to reopen the issue, and, on November 19, the provincial Financial Commissioner and Judicial Commissioner came to ask when the Dutch were going to leave. Van Hoorn replied that they could leave in a few days once the client-merchants had settled their debts. This had the desired effect; the debts were settled in a few days, and, on December 4, Van Hoorn took his leave of the officials and left Foochow for the ships. Net profits on the 1666 cargo had been only about 65 percent, not at all adequate when shipping expenses were allowed for, and the returns apparently were about half in silver because of strict prohibitions of the more desirable silk exports. None of the 1667 cargo had been sold. The ships set sail on December 17, and arrived at Batavia on 9 January 1668.[40]

At Batavia, the documents from Peking were received with much pomp, including musket volleys from the garrison and salutes from the cannon of the castle, and then opened and translated with growing puzzlement and exasperation. Van Hoorn submitted an excellent systematic report on the trade and manufactures of China, and recommended that the Dutch try to continue trading by sending two or three ships to Canton with a middle-sized cargo and some presents for the Emperor. He noted the ambiguities in the documents from Peking as translated, but thought the Dutch probably would be allowed to trade every year if they sent some presents to the Emperor each time, and that perhaps these presents could be forwarded by the Ch'ing authorities so that it would not be necessary for a Dutch envoy to go to Peking. Once again, we see how little even an intelligent Dutchman understood of the tribute system; the presents were important, but an ambassador always accompanied them; and, if an embassy was sent before the expiration of the statutory period, it would be turned away or perhaps accepted as a special case.[41]

The Company had spent about f132,000 on presents and expenses of the embassy; its profits on trade in 1666-1667, which would not have been allowed if the embassy had not been sent, were much more than that, but it is by no means certain that there was any real profit when we remember that five ships were used one year and three the next. And the effort and the expense had been a complete waste, or perhaps even counterproductive. For northern European knowledge of China, the gain was considerable. Young Joan van Hoorn rose to be Governor-General of the Company from 1704 to 1709, and made a modest contribution to the spread of knowledge of Chinese medicine; he had a Chinese doctor in his service for many years, and took him along when he went home to Holland in 1709.[42] The great compendium of Olfert Dapper, bringing together many important records of Dutch relations with China from 1662 to 1664 and on the Van Hoorn embassy and adding a great deal of information drawn from Jesuit sources, was published in the 1670s in Dutch, German, and English, in fine folio volumes with engravings of Chinese cities, the Peking palaces, and so on, taken from the drawings done by the artist in Van Hoorn's suite. It seems to have sold well—copies still turn up fairly regularly in antiquarian catalogues—and was summarized in many eighteenth-century collections of voyages.[43] This is the only one of our four embassies that was publicized in such fashion in its own time. Van Hoorn himself was the author of what may be the first original contribution in Dutch to the European idealization of Confucius, a long poem entitled "Some important qualities of true virtue, prudence, wisdom, and perfection, taken from the Chinese Confucius, and put into rhyme by the Honorable Pieter van Hoorn."[44] The knowledge of Confucius must have been drawn from Jesuit works. In much of the poem, the overlay of European ideas and modes of feeling is pretty thick, but occasionally Confucius really is there:

Many fear VIRTUE, lost in sloth and shadow,
Because it seems a mountain, far and hard to climb.
But Man! How can VIRTUE be so far?
If truly wish'd, then VIRTUE's near at hand.
. . . .
Rule then the people, deal with the common kind,
As if you as holy priest the Temple did approach.

There also are traces of the embassy in Chinese literature.[45] Wang Shih-chen, the most famous poet of the time, was a Board of Ceremonies official at the time. He was particularly impressed with the swords the Dutch brought, which were so flexible they could be bent in a circle, and by the horses and oxen, and wrote three poems. The sword poem is especially effective:

They call to mind the battle at Quemoy,
How the royal forces joined around the enemy.
Sent from beyond the sea, how chill their gleam!
Magic swords of legend to be sure.

Another poem is by Ch'en Wei-sung, later famous but still obscure at the time. His poem combines tall tales about the wealth of tropical lands that send such presents, a Confucian perception that even foreigners may be teachable, and astonishment at European wigs:

Fine curling hair that falls in even waves,
Precious swords that cut greens, cut silver.
Ceremonious in their customs, they can become tractable;
As a sign of respect they take off their hats.
. . . .

They bring amber big as cartwheels,
Coral ten feet long, in fresh colors,
Sandal and calambuc tall as a man;
Sometimes they chop it up for firewood.

chapter three

Manoel de Saldanha

1667-1670

Manoel de Saldanha arrived in Macao from Goa one day after Pieter van Hoorn left Peking. In the Ch'ing records, Saldanha's embassy, like Van Hoorn's, was a tribute embassy, and many details of their bureaucratic management were closely comparable.[1] But the study of the Saldanha embassy requires some different approaches, and leads to some surprising conclusions. The sources are more varied and problematic, with nothing like the consistency and day-to-day detail of the Van Hoorn day-registers. But there are some excellent Jesuit sources, of which the most important are extensive diaries by two eyewitnesses, Fathers Luis da Gama and Francisco Pimentel. Portuguese official records preserved in Goa and Lisbon are fragmentary but vital. The records of a Dutch attempt to trade at Canton in 1668-1669 fill in a few important details. The oddest source of all is a Ch'ing document recording the testimony of the merchants and minor officials involved in extortion from and illegal trade with Macao in 1667.

An understanding of this embassy also requires some knowledge of several contexts that affected it: the Portuguese empire in Asia; Macao, its trade and its local government; the Roman Catholic missions in China; the Ch'ing power structures in Canton and in Peking and their relations with Macao and the missionaries. In a nutshell, the embassy was sent to save Macao; the Peking Jesuits did much to save both the embassy and Macao; and they, in turn, could have done nothing if the K'ang-hsi Emperor had not wanted to favor them, and by extension the embassy and Macao, for his own domestic political reasons. Especially in following the whole sequence from the beginning of Macao's travail in 1662, often the only real clues to the reasons for the court's decisions are their *dates,* and other decisions the court made at the same time about the Dutch, the missions, or the high politics of the court itself. As shown in Chapter 2, such time linkages sometimes help to understand the court's decisions about the Dutch. The linkages seem to be even more striking here, despite the minuscule place of Macao's affairs in the vastness of the Ch'ing polity. Perhaps this was because the policy toward Macao was so closely linked to the standing of the Peking Jesuits, who, in turn, were totally dependent on inner-court politics and in particular on the favor of the Emperor. In this chapter, the embassy seems almost overshadowed by its contexts, but the results support the emphasis on contexts with which we began; if we started this story in 1667 and confined it to the embassy as such, we would have a much shallower understanding of its goals and its place in Ch'ing politics. For the Portuguese and the Jesuits and the Ch'ing authorities, the embassy was intimately connected with the travail and the survival of Macao from 1662 onward.

The results of a study of this embassy are a little surprising. This brave, confused, impoverished effort was the capstone of a process that confirmed the Portuguese position on the China coast at a time when all the wealth and intelligent management of the Dutch East India Company could accomplish nothing. The tribute embassy was essentially a court institution, with many links to the inner court, and friends at court could make up for many other weaknesses in a particular embassy.

MACAO IN PERIL

In the 1660s, Macao already was over a hundred years old, and was in the first of the many crises that make its history the supreme example of that refusal to give up a "lost cause" that contributed so much to the longevity of the Portuguese empire. Macao's trade had been shaky and occasionally overextended from the 1620s on, but the real setbacks had begun with the final absolute exclusion of the Portuguese from Japan in 1639, followed by the Japanese execution of the entire embassy sent to seek reconsideration of the ban in 1640.[2] In 1641, Malacca, key to the Portuguese trade route between Goa and Macao, fell to the Dutch, and, in 1642-1643, the Dutch took two Macao ships off Malacca and three more returning from Timor.[3] Even Portugal's regained independence from Spain, celebrated exuberantly in Macao in 1642, disrupted the important trade between Macao and Manila, but apparently not for many years.[4] In 1644, Peking fell to Li Tzu-ch'eng and then to the Ch'ing. For the rest of the 1640s, trade and manufactures were disrupted throughout China. There were dangerous pirates in the Canton estuary. In 1647, the Ch'ing General Li Ch'eng-tung took Canton by a ruse, but soon changed to the Ming loyalist side. The Ch'ing took Canton again on 21 November 1650, and a frightful massacre followed.[5] In much of China, 1648 was a famine year. Macao's food supplies dwindled, and 5,000 people, perhaps an eighth of the refugee-swollen population, starved to death.[6] No one could be sure what the harsh new rulers in Canton might make of the thin tissue of Ming precedents for toleration of this alien growth on Chinese soil, especially in view of the Christian connections of the Ming loyalist Yung-li court and Macao's having sent 300 soldiers and two cannon to aid that court.[7]

But Macao survived. In addition to the Macao-Manila trade, direct voyages to the Portuguese centers in India gave the Portuguese an advantageous position in supplying Chinese tutenag to the great Indian brass industry. A thriving trade grew up with Macassar and Timor in Indonesia, especially valuable because of the Chinese market for Timor sandalwood. The key figure in this development was the remarkable merchant-diplomat Francisco Vieira de Figueiredo.[8] His influence and this trade grew during the Dutch–Portuguese truce, in effect in Indian waters from 1644 to 1652, and survived the resumption of war in the 1650s. Even more important for Macao's survival were its relations with Chinese power-holders. In 1642 or 1643, it established cordial personal relations and even some commercial ties with the great sea lord Cheng Chih-lung.[9] In 1647, after the first Ch'ing conquest of Canton, Chinese merchants were allowed to come to Macao to trade, because the tariffs on the trade would augment the Ch'ing government's war funds, but Portuguese were not allowed to come to Canton. In January 1651, after the second conquest, the Governor of Kwangtung reported that the Westerners had been settled at Macao, trading and paying tolls, for over a hundred years. Now all had submitted to the new dynasty: "Truly this is because the Emperor's virtuous teaching has spread far and wide, and all far and near have submitted."[10] Despite the echoes of "tributary" rhetoric here, no mention was made in this memorial of any requirement that the Portuguese send an embassy to Peking, and there is no evidence that the Portuguese ever were told they should do so in order to regularize their status.

Many of the powerful Chinese who were favorable to Macao in these years were to some degree sympathetic to Roman Catholicism or at least friendly with missionaries. The Shun-chih Emperor himself had a warm personal relation with Schall, whose role in frustrating the De Goyer–De Keyser embassy already has been seen. Cheng Chih-lung had spent some of his early years in Macao, had been baptized there, and had many Christians in his entourage.[11] The Governor-General of Kwangtung and Kwangsi who first proposed in 1647 that the Portuguese be permitted to stay at Macao was T'ung Yang-chia, member of an immensely powerful Manchu-Chinese family, of whom many were patrons of the Jesuit missionaries and some converted to Roman Catholicism. In the 1650s, the Feudatory Prince Shang K'o-hsi and his son Shang Chih-hsin favored the Portuguese, and Chih-hsin, who had been friendly with Schall when he lived in Peking, treated missionaries and Portuguese as personal friends. The Shangs also had extensive commercial ties with the Portuguese, in a period characterized by the revival of some of the corruption of late Ming government and by the unusually frank and vigorous commercialism of the Manchu and Liaotung Chinese elites.[12] Contacts were mediated by Portuguese or mixed-blood interpreters like those who interpreted for the Dutch in 1652 and 1655,[13] and by Chinese intermediaries and client-merchants, some of them Christian. From the 1660s on, we can trace tentatively names and careers of some such individuals: Boneca ("The Doll"), Li Chih-feng or João Li Poming, "Barbarrão," and others.[14] The Portuguese may have found these ties and dependencies expensive and distasteful, but they linked the interests of the Canton officials to the survival of Macao. Also, once the Ch'ing Government had decided to tolerate Macao, they definitely considered it to be on Chinese territory, and so would not permit attacks on it by enemies of Portugal, as the Dutch were clearly informed in Fukien in 1662.[15] The Ch'ing officials also found Macao tolerable because it was so easy to control. Trade with China was its only source of income. All its food came from farms in Chinese territory, out beyond the "Circle Gate" on the narrow neck of land, where a small garrison kept watch and a sub-magistrate dealt with the Portuguese authorities and maintained control over the city's Chinese population, entering the city whenever they saw fit.[16] Macao usually could be brought to heel very quickly simply by keeping the gate closed. Portuguese institutions did little to impede Ch'ing control. The viceregal government in Goa paid little attention to Macao, Lisbon less, as any scholar knows who has tried to find documents on Macao in the general administrative archives of these governments.[17] Goa appointed the captain-general who commanded the garrison, but the government of the city and the management of its relations with the Chinese and its foreign trade were in the hands of an elected Senate of local Portuguese, who knew that good relations with the Ch'ing authorities were essential to Macao's survival. Captains-general sometimes protested the Senate's craven submission to the Chinese and unwillingness to pay the garrison, but to little effect.[18]

In 1662, the tentative and anxious survival of Macao was threatened by the consequences of the notorious Ch'ing coastal evacuation policy.[19] Frustrated in their efforts to attack the Cheng regime in its island bases and to cut off its channels of trade and food supply on the mainland, the Ch'ing Government began in 1660 to evacuate all the people and burn all the towns in a strip ten miles deep along the southern Fukien coast. In 1661, these measures were extended to the entire coastline of Chiang-nan, Chekiang, Fukien, and Kwangtung. In October 1661, two Manchu high officials were dispatched from Peking to Kwangtung to consult with Prince Shang K'o-hsi and the high provincial officials on coastal fortifications and evacuation measures.[20] The dispatch of Manchu officials to provincial capitals was a key mechanism of control and of enforcement of distasteful policies during the Oboi Regency.[21] These officials may have returned to Peking or sent in another report recommending full implementation. They must have had to overcome considerable resistance from Shang and the Kwangtung officials. According to an official Portuguese report,[22] on 3 May 1662, six lower-ranking officials arrived in Canton bringing news of the court's final decision for full implementation. In many coastal areas, according to the Portuguese, people were simply driven from their homes and left without food or means to buy it, so that many of them died, and others rose in futile rebellion and pillage. Later, the Ch'ing simply killed everyone they found in coastal areas; this may have been after a deadline expired or where they encountered resistance. Looting and robbery were widespread; according to a later Dutch report from Fukien, some captive women and children were enslaved and sold in Peking.[23]

According to the above-mentioned Portuguese report, some time between May 3 and 14, a fleet of war-junks under a maritime Brigade-General appeared off the beaches of Macao, and the General and his suite passed through the city to the border gate without telling anyone why they had come. According to Francisco Vieira de Figueiredo, who was not there but had close commercial relations with Macao, the official in charge of this party was "an Inspector General, who is like a God among the Chinese," and the Portuguese already had heard that they would be ordered to evacuate Macao, but had paid little attention, thinking that, since they were Portuguese, the evacuation orders would not apply to them. Vieira also wrote that, when the official had passed through the city, Chinese residents urged the city officials to go to the gate at once with a big present for the official, but they paid little attention, and the official, seeing they were not coming, raised the red flag, signifying that the gate was closed, and left.[24]

On May 14, a proclamation was posted in Macao ordering the evacuation of all its Chinese residents within three days. The Chinese hurried to obey, leaving behind a population of about

200 to 300 Portuguese and other Christian males, perhaps a nearly equal number of married Christian women, and about 2,000 widows and orphans dependent on church charity.[25] (The non-Portuguese Christian males probably included baptized Asian and African slaves and other dependents of the Portuguese; presumably Chinese Christians had to leave.) The city was deprived of all its craftsmen, boatmen, and small traders in food and other daily necessities. Our sources do not mention maritime trade, but clearly it was prohibited. The gate remained closed, and, since Macao grew little or no food, stocks must have been exhausted within a few days. A Macao delegation hurried to Canton. There Shang K'o-hsi "received them in public with asperity and in secret wept with them,"[26] because of his interest in trade with them and the contribution that trade had made to the prosperity of Kwangtung. According to Vieira,[27] Macao offered to pay half of Shang's expenses if he would use his influence in Peking on their behalf; a second-hand Jesuit chronicle says he was promised 40,000 taels if he would obtain permission for them to resume their maritime trade.[28] For big bribes, says another Jesuit source, some craftsmen and boatmen were allowed to return to Macao and a little trade was allowed "with the greatest stealth possible."[29] The officials agreed to "suspend or dissimulate" the orders cutting off food supplies, and to allow the sale to Macao, for silver, not in exchange for goods, of 30 piculs of rice per day, perhaps a pound per person per day. When the city authorities protested the small quantity, "they replied that if it was not enough they could go away."[30] Now the Macaenses might hope to escape immediate starvation, but had to go to the gate "in rain, cold, or calm, to seek what was needed every day."[31] Prices were so high that it seemed that "the Chinese sought this expedient to take all our silver, killing us and impoverishing us at the same time." Ch'ing soldiers were killed, their commanders took steps to bring them under control, and both sides settled down to wait.[32]

In keeping with the strong ecclesiastical hue of Macao government and society, one of the first responses of the city to the crisis of 1662 was a series of barefoot penitential processions of civil and ecclesiastical dignitaries and most of the populace, seeking divine assistance and the intercession of Saint Francis Xavier.[33] Another was to seek the help of the Jesuits in China, writing both to the Fathers in Peking and to Father Jacques Le Favre, the new Vice-Provincial. Early in the summer of 1662, Father Le Favre received a letter from Macao and set out from his post at Kan-chou, Kiangsi, for Peking. When he arrived, he found the Peking Jesuits already had been hard at work communicating with all at court with whom they were on good terms. They had learned that all they could hope to accomplish for the present was a decree to the Kwangtung authorities recommending that Macao not be destroyed; the next step would be decided on in consultations in Canton. The Peking Fathers decided Father Le Favre should go to Canton as quickly as possible, to be on hand while these consultations were taking place.

He set out early in the winter, enjoying official status as a carrier of documents for Schall. Thus, he was received with honor along the way and probably gained much face for the Jesuits and for Macao in Canton. But, after much discussion among the Canton officials and between them and the Macao civic leaders, Shang and the others said they knew the Regents were determined to cut off maritime trade, and would not recommend its restoration, but did include some praise of the city in their report to the throne. The Macao authorities then urged Le Favre to make haste back to Peking and work with the Peking Jesuits and the Macaenses residing there (possibly survivors of Cheng Chih-lung's household) to make the most of these few words of praise.

Le Favre did hurry, arriving in less than two months. Schall especially did all he could, risking much of the good will he had accumulated in his long years in Peking. The case of Macao was vehemently debated, says our Jesuit source, at the highest level of the government, and the reports from the Kwangtung rulers were read over and over. Finally it was resolved that the Kwangtung rulers be asked why, since they said Macao did not deserve to be destroyed, they recommended forbidding its maritime trade, which was the only way it could survive in its present location. This decision was referred to the Board of War, by whose recommendation trade had been cut off, and the Board recommended that Macao be depopulated, and all its foreign residents sent to their homelands. So now the Fathers began all over with prayers and masses and visits to their friends at court, and finally it was decided that these recommendations would not be implemented until the Kwangtung authorities had had a chance to explain their previous statement.

This decision-making process was as slow and subject to repeated reverses as that which so exasperated the Dutch at this time, but to the Jesuits and Macaenses every month of delay was a month of survival for Macao and more time for persuasions that might ultimately be successful. Finally, the court received more favorable recommendations from Canton, and the "Council of War" now voted without dissent to allow Macao to trade. The Fathers rejoiced, but then learned that the Regents had rejected this recommendation, and had ordered only that Macao be preserved, and not lack what was needed for the sustenance of life, but not be allowed to carry on maritime trade. The timing of this whole sequence is unclear beyond Le Favre's first trip to Canton early in 1663. If this final reversal came early in 1664, it came in the midst of uncertainties and divisions at court about relations with the Dutch. The Peking Jesuits later believed that the court had become uneasy about Macao, and later revoked the Dutch trade privileges, because of the frightful impression of European guns and warships brought back by court officials who had witnessed the Dutch-Cheng battles of November 1663.[34] The reversal on Macao was followed by another decision by the Regents that it was dangerous to confide control of such an important coastal strong point to foreigners, to whom the Ch'ing might be odious because of past aggravations and insults, and that all foreigners at Macao were to be sent away and all fortresses there razed so that they could be of no use to the Cheng forces or other maritime challengers. All this was decided in such great secrecy that the Fathers thought it a miracle that they had learned of it at all. This time they made especially good use of the influence of an official who was on good terms with Fathers Buglio and Magalhaens, possibly Hsu Chih-chien,[35] and, "by means especially apt to the Tartar nature," very possibly bribery, secured a revocation of the last decree. The former remaining in force, Macao was to be preserved but not allowed maritime trade. When this decision became known in Canton, the officials told the Macao authorities they should not despair, for soon the Cheng forces would surrender (this probably was a reference to the plans for expeditions against Taiwan in December 1664 and May 1665), and then maritime trade could be opened.[36]

This optimism was misplaced for a variety of reasons, none of them directly related to Macao. First, after the failure of the two Ch'ing expeditions against Taiwan, the court turned in 1665 to efforts to negotiate with the Cheng regime, keeping up the economic pressure by very severe enforcement of coastal evacuation and prohibition of maritime trade. New guard posts were established in the Canton Delta.[37] Thus, it was less likely than ever that Macao's maritime trade would be reopened, and the probability that it would be completely depopulated became somewhat greater. Second, in the spring of 1665, Yang Kuang-hsien's accusations led to the trial of Schall and his associates in Peking, the closing of all Catholic churches in the empire, and the assembly of all the missionaries in uncomfortable and contentious house arrest in Canton.[38] Macao was under a very black cloud because of its close association with the missionaries, and was deprived of its best channel of influence in Peking. Third, in 1665 and 1666, the Regents were taking one small step after another to undermine the power of the Three Feudatories in the south, including Shang K'o-hsi, Macao's best friend.[39] The political situation became even less favorable to foreigners as Oboi sought dictatorial power in late 1666 and early 1667; here again, Macao's experience offers a startlingly close parallel to that of the Dutch.

We have only fragments of information on the situation at Macao between the middle of 1662 and the middle of 1666. The Tanka of the Canton Delta rose in resistance to the coastal evacuation policy in 1662, and one Portuguese document hints of Macao aid to the Ch'ing against them.[40] In any case, the Tanka rebels and the Cheng squadrons that occasionally cruised along the coast would have made the Ch'ing think twice before gratuitously eliminating a foreign strong point that might divert enemy hostility from Ch'ing areas and whose forces might even be used actively against a seagoing enemy, as the Dutch had been in Fukien. In 1662, three inbound Portuguese ships were lost in a storm.[41] Arriving ships were not allowed to come into the harbor, but two stayed out in the islands and managed to slip away on the north monsoon.[42] In June and July 1663, two ships arrived; the commanders of one, belonging to Francisco Vieira de Figueiredo, had not known of the prohibition of maritime trade. The Macao authorities sent 5,000 taels by way of "General Li," probably Li Chih-feng, and 3,000 by way of Boneca, to secure the release of these ships, and probably also hoped by these bribes to obtain permission to sell their cargoes to Chinese merchants. Five more ships arrived in 1664. Their cargoes were confiscated by the Ch'ing authorities, but then "General Li" and the others told the Macaenses they could have their cargoes back for a payment of 13,000 taels to the Governor-General.[43] Macao paid, and the cargoes were returned to molder in Macao's warehouses until 1667. One ship slipped away or was allowed to leave in December 1664.[44] Most of the others, and perhaps five or six more that arrived in 1665,[45] seem to have stayed in waters near Macao, more or less closely guarded by Ch'ing war-junks, slowly smuggling their cargoes ashore past the Ch'ing junks. The Canton authorities and/or their merchant intermediaries extorted another 4,000-5,000 taels in silver, jewels, and import goods some time in 1665.[46]

Our knowledge of the course of events improves abruptly in November 1665, the first surviving date in the excellent diary of Luis da Gama, S.J.[47] The primary authors of the extortions of 1665-1667 were two Governors-General named Lu. Both were Liaotung Chinese. Lu Hsing-tsu, who had been Governor, succeeded Lu Ch'ung-chün early in 1666; Da Gama refers to both by their office, and I can detect no differences in their actions toward Macao.[48] Shang K'o-hsi, like Keng Chi-mao at Foochow, had little influence on relations with foreigners in these years. The leading intermediaries between the Governors-General and the Macao authorities were the various client-merchants and the Magistrate of Hsiang-shan, the sub-prefecture that included Macao, Yao Ch'i-sheng.[49]

In a time when many entered the bureaucracy and rose and fell in it in eccentric ways, few rode as dizzying a roller coaster as Yao Ch'i-sheng. A late Ming *sheng-yuan* from Chekiang, he was fond of the martial arts. When the Ch'ing conquered the Yangtze Valley, he probably was leading some kind of "knight-errant" gang in T'ung-chou on the north side of the estuary. He came into conflict with local "bad gentry," surrendered to the Ch'ing, was made Magistrate of T'ung-chou, but was dismissed after he had one of his "bad gentry" antagonists seized and beaten to death. Returning to his knight-errant ways, he killed two Manchu soldiers, returned a girl whom they had seized to her family, and, attaching himself to the family of a kinsman, enrolled in the Chinese Bordered Red Banner. He became a provincial graduate in 1663 and was appointed magistrate of Hsiang-shan; in 1668 he was dismissed for the violations of coastal prohibitions which we will follow in detail below. But he was not done, not with knight-errantry, nor with bureaucratic fancy footwork, nor with dealing with foreigners. In 1674, when Keng Ching-chung rebelled in Fukien and advanced into Chekiang, Yao and his son raised a private army and joined the Manchu commander, Prince Giyešu. He reportedly contributed large sums of money as well as his private army to the Ch'ing cause. Giyešu was famous for spotting talented men and advancing them rapidly; in 1678, Yao was named Governor-General of Chekiang and Fukien. He held this post until his death in 1684, dealing with the Dutch traders, negotiating to split and undermine the Cheng regime, and providing organizational and logistical support for Shih Lang's conquest of Taiwan in 1683.[50]

Late in November 1665, Yao and General Li came to Macao with an audacious proposal from the Governor-General: for 100,000 taels, he would procure from the court full permission for Macao to reopen its maritime trade. The Portuguese figured that, together with the sums they already had paid and 2,000 taels for each ship, presumably the measurement dues for ships already on hand, they would have to pay about 150,000 taels to be allowed to resume their sea voyages. Yao and Li said 70,000 of the 100,000 taels could be paid when trade was opened, but the Portuguese would have to pay 30,000 in advance.[51] (In a similar situation, Ch'ing officials once explained to the Dutch that they had to have cash in advance to be able to make the necessary bribes all the way up to Peking.)[52] The Portuguese had no such quantity of money on hand, but they turned over goods worth 12,000-15,000 taels as a pledge that the 30,000 would be paid. If they had finally had to pay the full 100,000 taels, it would have been a heavy burden on their rather small volume of trade. But they could hardly help going along; if the officials did not get what they demanded, they could extinguish Macao any day between morning gruel and noon rice, and, if they finally did obtain a relaxation of the trade restrictions, Macao could at least begin to pull its trade network together again while it struggled to pay what it had promised.

In the spring of 1666, it was not at all clear if this latest payment would bring any benefit to Macao. The Ch'ing officials said the ships that had arrived from various quarters in 1665 would have to move to a distant anchorage out of sight of Macao; probably they were expecting official visitors from Peking. In June, the Ch'ing war-junks that had been guarding the Portuguese ships moved away from them and they were able to send some of their cargo ashore. But the south monsoon had set in, and it was too late for them to load more cargo and leave; only one small junk left Macao that spring. There were several rumors that the Chinese merchants were about to bring boatloads of goods from Canton and buy Portuguese import goods, but nothing happened. One report had it that the boats had been loaded and ready to go, but at the last minute Shang K'o-hsi refused to let them go, saying he did not want to lose his head. But, in September and October, the war-junks stayed away, the Portuguese ships came back to the anchorage near Macao, and the Portuguese gained a bit of favor by sending soldiers in boats to attack and defeat pirates near Shang-ch'uan Island.[53]

Early in November, the people of Macao were astonished by the arrival in the waters around them of a fleet of 60 to 70 war-junks carrying 5,000 or 6,000 soldiers. When the city authorities went to the officers in charge with gifts, the officers refused the gifts and told them that the Portuguese ships must leave immediately or be burned; there must be no sign of them left, for fifteen inspectors had arrived from Peking to check over the coastal areas. (Several references to capital officials in Foochow and adjacent coastal areas late in 1666 and early in 1667 were noted in Chapter 2.) In a later interview, they were given a deadline of three days. Not all the ships could be ready to put out to sea that soon; some were old and perhaps unsound, and others may have been undergoing major repairs. Some in the city favored requesting extension of the deadline, and going to seek it with a small show of force in armed boats, but the majority favored acquiescence and so informed the officials. The officials were pleased and suggested that the Portuguese burn the ships themselves, and, on the night of November 14-15, they did so. Four ships and three junks were burned. One other sank so that only the poop and masts were showing, but, when the Chinese went out to finish burning it, the owners bought them off for 1,500 taels, and the Chinese even helped the Portuguese refloat it, so that it was ready to set sail in five or six days.[54]

What happened here? The Chinese clearly were just as happy to see the ships leave as to see them burned, and very probably could have been persuaded or bribed to give the Portuguese a few more days to make all the vessels seaworthy. Was the Macao leadership so demoralized it couldn't think straight? Or were some of the shipowners thinking they might be able to hold the city government or even the Canton authorities liable for the loss of ships that already were aging and rotten?

A ship set sail for Goa on the night of November 14, in full view of the Ch'ing war-junks, carrying a letter from the Senate of Macao to the Viceroy and Council in Goa.[55] They had repeatedly asked for subsidies and reinforcements for the garrison, wrote the Senate, to no avail, and now they held on to Macao solely out of loyalty to the Portuguese Crown. They recommended that an embassy be sent to the court in Peking in the name of the Portuguese King, and, if Goa did not send an embassy, it should send ships with enough space in them to allow the people of Macao to board them and abandon the city once and for all. It seems likely that the Portuguese were seriously considering leaving. Many people were starving. The Senate and the ecclesiastical authorities were especially concerned about the large number of women dependent on charity; Domingo Fernandez Navarrete, O.P., wrote that "of late years many women expos'd their Bodies to Infidels for Bread."[56]

In December 1666, the Portuguese learned of rumors in Canton that the court was about to allow Canton merchants to come to Macao to buy Portuguese import cargoes. But then they heard that the four Regents had reversed bureaucratic recommendations favoring these permissions, and had ordered that, since Macao was exposed to pirate attacks, it no longer should be exempted from the coastal evacuation orders, but all its people should be moved back within the evacuation boundaries.[57] This reversal by the Regents, known in Canton in December 1666 or January 1667, came at very nearly the same time as the Regents' reversal of ministerial recommendations on the Dutch. In both cases special privileges for foreigners were canceled; in both, although on very different scales, the foreigners had been used against pirates, and might have been persuaded to be so used again. In both, as discussed in Chapter 2, the rise of the Oboi faction and the fall of Sunahai probably were the most important causes of the reversals.

On 15 February 1667, Yao Ch'i-sheng came to Macao to formally present the imperial order to move inland. He suggested some arguments for the Senate to use in its protest, which would be transmitted to Peking by the Canton authorities, and then suggested that Macao promise 250,000 taels for the opening of its maritime trade. Lu Hsing-tsu, it soon became clear, would be the main mover in this effort to persuade the court to exempt Macao from the most basic of the coastal prohibitions, and the 250,000 taels would reimburse him for bribes up and down the bureaucracy and repay him amply for his trouble and risk. The Macao authorities readily agreed to Yao's proposal, saying to themselves that they would have promised 400,000 to keep the gate open for food supplies and thus to keep from starving until ships came in from Macassar and India with food supplies and other reinforcement in June or July.[58]

Once the Macao authorities had agreed, Lu seemed to keep his side of the bargain. He wrote to Peking on Macao's behalf; a reply was received in the middle of April. Also, from February 18-19 to March 14-15, the gate was opened every five days. But thereafter it stayed closed until April 29, amid signs of the Canton officials' intense anxiety that their toleration of Macao and of some of the actions of the Portuguese there might cost them their offices or their heads. It is not clear what started this change early in March; by mid-April, the Canton officials probably knew that Chang Ch'ao-lin was in big trouble for small favors to the Dutch in Foochow, and certainly knew that three high officials had been executed in the Peking area for protesting the impact on the common people of Oboi's manipulations of Banner landholdings.[59] The officials were extremely worried about a Portuguese ship and a junk that were anchored in the outer islands; they questioned the Macao authorities repeatedly about them, but the Macaenses denied that the craft were Portuguese. The Ch'ing officials sent a squadron of junks out to check on them, and there were repeated rumors that a larger squadron would go out to drive them away, until they finally were blown away from their anchorages in spring storms and did not reappear. From March 21 on, proclamations were posted in Canton and sent to Macao that the people of Macao were to be moved inland without further delay, that a place had been prepared for them to live, and that they should be all packed up and ready to go. These orders continued even after the officials had resumed negotiations with Macao, and both the orders and the questions and threats about the vessels in the outer islands probably were intended largely to avoid trouble with any investigators who might come from Peking.

Far more peculiar, and it seems to me even more patently an effort to make appearances right in anticipation of an investigation, were the officials' efforts to obtain the cargoes of the two ships that had arrived at Macao in 1663, later expanded to include those of the five 1664 arrivals. Some of these goods had arrived in bad condition; most of the rest had since decayed into nearly worthless powder. Some had been bought by Canton merchants but left in Macao warehouses as Ch'ing restrictions tightened. In the Ch'ing view, all was contraband, having reached Macao—

Chinese territory—in violation of the maritime trade prohibitions, and all should have been confiscated long ago. Early in 1667, Peking somehow learned that these cargoes were still in Portuguese hands. The goods would have to be in official warehouses in Canton to show to an inspector. On March 15, officials came to demand on behalf of Lu Hsing-tsu the surrender of the 1663 cargoes. The Senate explained that payments had been made via Li Chih-feng to be allowed to keep these cargoes; they subsequently had to exert themselves for several months to explain that other merchant intermediaries also had been involved in the earlier payments and to get Li out of jail and back in the good graces of the Canton authorities.

Yao Ch'i-sheng came to Macao on April 21–22 with a very mixed bag of news. The court had ordered, in reply to Lu Hsing-tsu's February report, that, if Macao wished to be exempted from evacuation, it should so inform Lu, who would memorialize on its behalf, presumably adding his own recommendation. Lu was pleased with the Portuguese agreement to pay 250,000 taels for the opening of its maritime trade, but conditions had changed since his February report; now, in view of the fate of the northern officials who had protested Oboi's land policies he was not willing to propose such a major policy shift. Peking also had ordered that the cargoes of the 1663 and 1664 arrivals be turned over as contraband. The gate would not be opened until they were. Nor, apparently, was Lu willing to add his positive recommendation to Macao's request to be exempted from coastal evacuation until these cargoes were turned over.

Instead of simply demanding that the goods be turned over, Yao now proposed, and Macao reluctantly accepted, a scheme by which the Chinese merchants who had bought Portuguese imports would be able to fetch them from Macao but would have to turn over four-tenths of these goods to the authorities.[60] These merchants and others also would be allowed to take goods to Macao to sell, buy Portuguese imports there, and sell these goods in Canton. The authorities would take stiff percentages of the goods at every stage and add these levies to the goods previously sent as bribes via Li Chih-feng and others, to make an ample store of "confiscated cargoes." In these negotiations and others, Yao often was less the remote magistrate who summoned people to do his bidding than a persuasive and accommodating mediator. On the evening of April 21, he decided to seek the help of the Jesuit Procurator, Father Luis da Gama (the author of the main source on all this), and, learning that the Father was sick in bed, spent several hours at his bedside in the Jesuit residence working out a deal.

After this agreement was made, the gate was opened once, on April 29, with an abundance of food for sale, then was not opened again. On May 17, Yao came to report that Lu Hsing-tsu approved of the April 22 agreement, but demanded 30,000 taels at once if he were to open the gate, withdraw the watch-junks that ringed Macao, and report favorably to the court on Macao's desire to be exempted from coastal evacuation. The city authorities insisted they could give only 20,000; Lu finally accepted this, and the Macao authorities went to work raising it by a general levy. On June 16, merchants came on seven big junks to trade, the first large, non-clandestine trade in more than five years. From this time on, the gate again was opened regularly, and adequate supplies of food were available. Two Imperial Commissioners were reported in Hsiang-shan, but somehow Lu and Yao persuaded them not to visit Macao. It probably was also late in June that Lu, honest extortioner that he was, wrote to Peking in favor of Macao's exemption from coastal evacuation.

The trade, of course, was in flagrant violation of the prohibition of maritime trade, and Yao had warned the Portuguese that Shang K'o-hsi and his merchant agents must not learn what was going on. The risk of favors to Macao also was made clear by the echoes of anti-missionary accusations of sedition and cultural corruption that appeared in the proclamations ordering the Macaenses to move inland. Van Hoorn in 1667 found many officials wary of receiving free-will gifts from foreigners; the extortion of massive bribes in return for specific policy recommendations was a much more grave violation of the norms of the bureaucracy than the receiving of such gifts.[61] Why was Lu willing to take such risks? Perhaps the sum that had been reported to Peking for the 1663 and 1664 cargoes was too large to be made up by simple confiscation of the goods in Macao. Extra profits from trade also would facilitate buying off any inspectors who came from Peking. Sheer greed for tropical imports and the profits of his client-merchants' trade no doubt was the most important motive. "Bringing back the confiscated cargoes" would give a thin veneer of legitimacy to the sending of merchants and junks to Macao. Greed is the only conceivable explanation for the extortion of bribes and promises of future payment. Lu also must have calculated that his court connections were so strong that he could ride out any storm resulting from all this. Perhaps he was close to the Oboi faction; the brief biographical materials on him offer no clues. Perhaps, by the time Yao reported to him Macao's offer of 20,000 taels, he had learned who the inspectors were who were coming to Kwangtung, and was sure they would make no trouble for him. In any case, his influence, and perhaps his bribes, did secure a full exemption of Macao from coastal evacuation, proclaimed and exultantly celebrated there on August 26.

But Lu had reckoned without the precocious, curious, and willful fourteen-year-old who sat on the Dragon Throne, and who passed a pleasant morning that summer viewing the exotic horses and oxen brought by the Van Hoorn embassy. On 25 August 1667, the Emperor proclaimed his personal assumption of ruling powers.[62]

Shang K'o-hsi, who had been excluded from all the negotiations and trade with Macao but who must have known most of what was going on, seems to have seen in this sudden change in Peking, which would have been known in Canton by late September, his chance to strike back at Lu Hsing-tsu and restore some of his own waning power. On the night of October 12, four low-ranking mandarins came in secret to Macao with a written order from Shang to the city to write up their grievances against Lu. The Senate refused, not wanting to get entangled in a Chinese legal proceeding. Lu then apparently imprisoned Yao Ch'i-sheng and tried to blame him for everything; on October 22, the Assistant Magistrate of Hsiang-shan came to Macao to take depositions against Yao. The Senate, recalling how much it was indebted to his mediation and good advice, said that, on the contrary, it would do all it could in his favor, "even if it came to giving blood from our arms in exchange for him." None of the maneuvering was of any help to Lu; Yao and the merchants involved made detailed confessions implicating him. Imperial Commissioners were sent to draw up a bill of particulars on Lu's crimes. He committed suicide in prison on 9 January 1668, probably after learning that he had been sentenced to death. Yao was dismissed from office, and he and the merchants lost much in fines and forfeitures, but, on the day of Lu's suicide, were pardoned for their capital offenses. About the same time, the popular Governor of Kwangtung, Wang Lai-jen, committed suicide in prison in an unrelated intrigue, leaving a posthumous memorial recommending the general revocation of the coastal evacuation laws and specifically calling for the preservation of Macao. By the end of 1668, the evacuation laws had been revoked, and commissioners were touring the Kwangtung coastal areas to supervise their re-population. The gate was opened every five days with few exceptions. Macao was not exempted from the continuing prohibition of maritime trade, but its enforcement seems to have been considerably relaxed. Chinese merchants came and went in Macao, often at night, and Portuguese ships arrived from overseas and departed again. There were more reports of memorials to the throne recommending the extinction of Macao in the summer of 1670, but by then the Jesuits were back in favor, Manoel de Saldanha was in Peking, and the Emperor had read the dossier on the appalling extortions of Lu Hsing-tsu.[63] Lu had served his victims well, in securing their exemption from coastal evacuation, in allowing them to survive the perilous spring of 1667, and in so overreaching himself that the Emperor came to see Macao as a victim deserving some restitution. The more sympathetic Shang K'o-hsi regained influence in Canton, and his own career of extortion was cut short.

A LONG STAY IN CANTON

Macao's urgent plea for an embassy, food, munitions, and reinforcements for the garrison, dispatched as the Macaenses burned their own ships on the night of 14 November 1666, reached

Goa about the end of February 1667. On March 14, the Council of State in Goa voted unanimously to send an ambassador, 100 Portuguese and 100 Indian soldiers, and supplies of rice and salt. On April 27, they added sulphur and saltpeter (both for gunpowder) and lumber, and reduced the number of troops to 80. Goa supplied neither cash nor presents for the embassy. All it did was send a letter to the King of Siam, asking him to continue to advance funds to Macao, and promising that the Goa authorities would reimburse him.[64] The King did advance more funds, apparently not worried about this contribution to the survival of a competitor in trade between China and Southeast Asia. For Macao, this financing was a mixed blessing; the city already had written to Goa to complain about its captains-general borrowing from the King of Siam without the concurrence of the Senate, and then leaving the city saddled with the debt when their terms of office were up; Goa had ordered that in the future the consent of the Senate would be required for all such borrowing.[65] Macao's debts to the King of Siam were not completely discharged until the 1720s.[66] Goa also sent a new Captain-General, Alvaro da Silva, to Macao in this monsoon season.[67]

But who was to go as ambassador? The tedium and lack of private profit made this an unattractive assignment even among the Dutch, and the financing and organization of this embassy would be far less adequate than for Van Hoorn's. But finally the Goa authorities found a volunteer, Manoel de Saldanha. Saldanha may not have been eager to go, but probably saw this as an opportunity to save himself from disgrace and exile. In 1657, he had been made the scapegoat for the loss of a border town to the Spanish amid general confusion and incompetence and had been exiled to India for life. He arrived in 1662, but had found no military employment that might have given him an opportunity to redeem himself.[68]

Saldanha sailed from Goa for Macao late in May or early in June 1667. He had been gravely ill, would be ill for much of the next three years, and finally would die before he left China. He probably was suffering from chronic malaria, complicated by varicose ulcers in one leg and by a least one bout of severe dysentery.[69] He was accompanied by his chaplain, Father Simão da Graça, and by Andre Gomes, O.F.M., who had come from Macao late in 1666.[70] He bore a letter for the Emperor of China written in the name of the King of Portugal, dated 12 March 1666, as if it had been sent from Lisbon, and sealed with the royal seal. But the Goa authorities had a royal seal to be used on correspondence with Asian potentates. There is no trace in surviving Goa-Lisbon correspondence of any discussion that could have led to the dispatch of this letter from Lisbon; indeed there is no record of discussion at Goa of an embassy to China before 1667. The Prince Regent approved the embassy after the fact, in 1670. The letter congratulated the Emperor on the great fame he had achieved at an early age and on his success in conquering the entire Empire of China. It is not clear if the drafters knew that the recipient would be the second Ch'ing emperor to rule in Peking, not the first; but it seems to have been fairly easy for sympathetic Jesuit translators to turn these phrases into the Chinese phrases appropriate to a "memorial of congratulation" accompanying the first embassy to a new ruler. It also asked the Emperor to "listen to the said Ambassador, and give complete faith and credit to what he may propose on my behalf"; but substantive requests were to be worked out in Macao.[71]

Saldanha spent a few days ashore at Malacca at the end of June, given lodgings by the Dutch East India Company authorities at their garden outside the town.[72] He arrived at Macao on 6 August 1667. Then nothing at all happened for several weeks. Saldanha's illness may have contributed to this as to many later delays. The Canton authorities must have learned of the embassy almost immediately, but may have decided to do nothing about it until they learned, late in August, the court's response to Lu Hsing-tsu's memorial in Macao's favor. Saldanha no doubt gained good will for his embassy and for Macao on August 31, when a brawl broke out between Portuguese sailors and the crews of the Ch'ing watch-junks, and he interceded to stop an assault on the junks by two companies of Macao soldiers; the Ch'ing local authorities then withdrew most of their junks, and Portuguese were able to begin unloading their import cargoes.

It probably was about the end of August, after the proclamation of Macao's exemption from coastal evacuation on August 26, and possibly after the August 31 fracas, that the Canton authorities sent a first report to Peking on the arrival of the embassy; the first signs in Kwangtung of any response from the court were around November 1. But the Canton authorities would have to see the Ambassador, his credentials, and his present and report on them to Peking, sending along a copy of the Portuguese King's "tributary memorial," before the court would permit the embassy to begin its long journey north. According to one source,[73] Lu Hsing-tsu said the embassy was not really from the King of Portugal but was a sham organized in Macao; before or during his imprisonment, he would have wanted to do everything he could to deprive the Portuguese of an opportunity to complain directly to the court about his extortions. After Lu's suicide the court and the Canton officials continued to insist on and finally received reasonable evidence that the embassy had not originated in Macao, but, in meeting the other reporting requirements and then sending the embassy on to Peking, they tolerated long delays, in striking contrast to the anxiety of the Foochow officials to get the Van Hoorn embassy started north. The immediate causes of the delays were the poverty of the embassy and the punctilio of the Ambassador, but I suspect the long delays suited everyone fairly well. It probably was not until early in 1669 that the Emperor began to be really eager, largely for his own political reasons, to see the embassy in Peking. And as long as the embassy was in Canton and was not in disfavor with the court, Portuguese ships could be allowed to come up to Canton to trade, on the pretext of bringing supplies for the embassy; probably there was less risk of official interference or confiscation in this way than if Chinese merchants went to Macao.

On September 9, a military officer came with five war-junks and twelve smaller boats to accompany the embassy up to Canton, the normal point of arrival and reception for tributary embassies from Siam and elsewhere, but he did not press the matter when the Portuguese told him that Macao was assigned as a special port for Portuguese ships, which did not go to the same port as Siamese and other ships. Then he asked for the letters that the Ambassador brought to the Emperor and to the high officials in Canton. The Ambassador replied that he would present "hand to hand" the "King's letter" to the Emperor, but would turn over the Goa viceroy's letters to the Canton officials. These were taken "to the sound of trumpets and drums" to the Senate House and ceremoniously presented to the officer. On September 29, more war-junks appeared near Macao, and the Portuguese learned that none of the four ships that had come from Goa would be allowed to leave until the embassy left, "because, it is said, the great mandarins of Canton fear that they would be deprived of their offices, as was done to those in Fukien, because they allowed some Dutch ships in which their embassy to Peking came and went, to arrive and depart." On October 6, a "Taquessy" (*t'i-chü-shih*), here a subordinate official of the vestigial Superintendency of Merchant Shipping, came to demand the royal letter and to be shown the presents to the Emperor and the cargo of the four ships from Goa. The City authorities gave him information about the cargoes, and said the presents (which very probably had not yet been prepared or even decided on) and the letter need not be shown until they were delivered in Peking. On the 11th, a higher "Taquessy" came to say that the Ambassador must come to Canton at once, and that the Canton authorities could not report to the court on the embassy until they had actually seen the Ambassador. But the Ambassador was sick in bed and unwilling to leave Macao in any case, and, on October 14, the official had to settle for a formal interview with Saldanha, who was still in bed, but had his room richly decorated, received him in the company of the Loyal Senate and many leading citizens, and showed him "two big trunks covered with gold-bordered carpets, in which were the presents for the Emperor" (or part of them, or perhaps none of them). But the Canton officials still were not satisfied and sent a *hai-tao* (maritime defense *taotai*) to persuade the Ambassador. Finally he yielded to the arguments of the city officials and the Captain-General and, on November 21, he set out for Canton on one of the *hai-tao*'s junks, accompanied to his embarkation by the city officials, much of the citizenry, and five companies of musketeers, and was sent off with salutes from the cannon of all the forts and the ships in the harbor.[74]

In Canton, Saldanha did not visit the high officials at once, and possibly not until March 1669. He continued to refuse to turn over the royal letter, until finally the Jesuits worked out a compromise that allowed the officials to report its contents to the court. The presents were not shown to the officials until December 1668, and there were efforts to improve them even after that. The embassy did not leave Canton for Peking until January 1670, over two years after its arrival in Canton. Many of the difficulties and delays of this period can be traced to the incredibly ramshackle financing of the embassy, some of which can be reconstructed from a detailed account of Macao's expenditures on the embassy and from letters—often rambling and vague—from Saldanha to various Jesuits in Macao.[75] The viceregal government in Goa contributed little or nothing to the cost of the embassy; even the hiring of the ship that brought the Ambassador from Goa and the price of the portrait of the King painted in Goa are included in Macao's list of expenses. Almost all the presents apparently were bought in Macao, although many of them were of Indian origin and would have cost much less in Goa. The Ambassador had a suite of 90, which had to be fed—sometimes very meagerly—for over two years in Canton. Portuguese sources do not mention any subsidy from the Ch'ing officials to the embassy while it was in Canton, but it is possible that it received rent-free lodgings and meager rations for 22 or 26 people. And Saldanha spent over one-fourth of the total expenditures of the embassy on fancy dress for his entire suite; a sensible form of ostentation for the 22 who finally were allowed to go to Peking, but almost a complete waste for the rest. It is not clear how the impoverished City found these funds; Saldanha's letters show him trying to borrow from Shang K'o-hsi and finding the City's credit nearly exhausted, working out complex arrangements with the City authorities on the disposition of Portuguese import cargoes, and melting down his table silver to keep his suite alive.[76] The money borrowed from the King of Siam did not arrive until July 1669.[77]

It also is clear that the Canton officials contributed to this long delay by meeting the usual reporting requirements not all in one memorial as they did for most embassies, including the Dutch, but one at a time; first the royal letter, then the presents, finally the Ambassador's credentials. The court in Peking tolerated or perhaps even prompted this singular procedure. As we shall see, it was not until the middle of 1669 that the political climate in Peking was really favorable for a Jesuit-connected embassy. The Canton authorities, to the degree that they wanted the embassy to succeed, would have sought to delay its arrival in the capital in the hope that prospects for a good reception would improve, while at court neither sympathizers nor opponents would have been eager to introduce another issue into an already explosive political situation. Also, the presence of the embassy in Canton seems to have provided a degree of legal sanction for a good deal of Canton-Macao trade in which the officials were interested.

After his arrival, Saldanha decided not to pay a formal visit to Shang K'o-hsi because he learned that, once such visits were begun, they would have to be made to all the high officials in turn, leading to "unbearable expenses," especially when they had to give banquets for the officials and when they were invited to theatricals at which they were expected to give substantial presents to the performers.[78] Apparently, however, the officials were willing to report to Peking on the embassy now that it was in Canton, even though they had not seen the Ambassador.

Saldanha's continuing refusal to turn over the royal letter was the next sticking point. After two or three exchanges of communications with the court on the matter, officials were sent from Peking to investigate the whole situation. Late in May 1668, they visited Saldanha, summoning Feliciano Pacheco and other Jesuits to inform them of the contents of the royal letter. They were very cordial to the Fathers and kowtowed to the royal letter when it was placed on a table. The officials then asked the Ambassador and the Jesuits to open the letter so that the Jesuits could translate it. The Ambassador refused, and the Fathers pointed out the royal seal that sealed the pouch containing the letter to the Emperor, suggesting that it would not be given full credit at court if that seal had been broken. I suspect the Jesuits knew that royal seal had been affixed, and the letter signed in the King's name, in Goa, but I cannot be sure. In any case, they were testifying to the genuineness of the *seal,* and can hardly be blamed for not volunteering information that would have scuttled the embassy. On the same day in May, lesser officials made inquiries in Macao about the authenticity of the embassy and also about the cargoes of the five ships that had arrived in 1667; these would be reported to Peking as the cargoes of the ships that had brought the Ambassador. All these investigations apparently led to a report to Peking that the embassy seemed to be genuine; when the Dutch arrived in the area in September, they heard that the embassy had been accepted and soon would leave for Peking. But it seems more likely that the court would not finally approve the departure of the embassy until it knew the content of the royal letter.[79] It was probably in September 1668 that Governor-General Chou Yu-te summoned Fathers Pacheco and Francesco Brancati. They had arranged to receive from Saldanha a copy of the text of the royal letter, and Saldanha had authorized them to translate it as they saw fit; they would send a copy of their translation to the Peking Jesuits, who thus could agree with the Canton version if called on to translate from the Portuguese original after it was opened in Peking. The Fathers found one sentence they feared would cause grave difficulties, and after much discussion with the Ambassador left it out of their translation. It simply said that the King hoped the Emperor would give full credence to Saldanha in all the business he would discuss. The Fathers told Saldanha that, if this sentence was translated, the court would insist on knowing in advance what business he planned to discuss, and if it concerned Macao he never would be admitted at all. When the text of the letter was reported to Peking, the court apparently replied that now the embassy's presents would have to be reported to Peking.[80]

While all these communications were going back and forth in 1668, parallel events were easing the plight of Macao and beginning to improve the climate for the reception of a Portuguese embassy in Peking. In June 1668, officials arrived from Peking to inspect the coastal areas. According to a proclamation at Macao in July, which was received with great jubilation, the Emperor had decided to permit the re-population of the mainland coastal areas, but not the islands. These officials had been sent to consult with the provincial officials and determine the specific areas to be re-populated and the garrisons to be established in them, a task that took the rest of 1668 and that often led to delays in other business when the high provincial officials were off touring coastal areas. On June 24, one of the Peking officials visited Macao, and, although he had been expected to tour the whole city, he went only to the Jesuits' Church of St. Paul and their residence, admired the organ in the church, said he had known Schall well in Peking, and, when Schall's portrait was brought out, he and his suite showed their pleasure at seeing it and their sorrow at his death. More watch-junks were withdrawn from around Macao just before the official arrived there, and, on the same day the missionaries in Canton were summoned to Shang K'o-hsi's palace, treated very cordially, and thereafter their confinement was considerably relaxed; the sympathy to Christianity of the official who went to Macao must have been one of a number of signs that convinced Shang that it was safe to show his friendship for them more openly.[81]

Saldanha and/or the Senate of Macao may have offered a bribe to the officials to insure that Macao would be regarded as mainland and not an island under the terms of the imperial decision, and, at the end of the year, the officials may have told Macao they would obtain full revocation of restrictions on Macao, including those on its maritime trade, for 120,000 taels; but nothing seems to have come of either discussion. Saldanha spent much of his time late in 1668 trying to improve the financial situation of his embassy, urging the citizens of Macao to pay the statutory "percent" taxes on their trade, and arranging for a Portuguese vessel that had arrived from Timor to make a voyage to Siam, bring back part of the promised loan from the King of Siam, and (apparently) make a large contribution to the embassy from the profits of its trade. All this was being done amid growing antipathy between Saldanha and some prominent people in

Macao. The Senate had sent with Saldanha two representatives to keep an eye on the embassy: Bento Pereira de Faria, who was supposed to serve as secretary of the embassy, and Vasco Barboza de Mello. They did not live with the Ambassador's suite, and he tried to exclude them from communications with the City concerning the second bribe negotiations mentioned above; since the Senate *was* involved in these negotiations, it may be that the faction that had controlled the Senate in 1667 no longer did in 1668, but that those now in control did not think it necessary or feasible to replace the embassy watchdogs appointed by their opponents. In any case, Saldanha sometimes appealed to the Captain-General in his discussions with the Macao authorities, but his usual tactic was to communicate via the Macao Jesuits and to give them a good deal of discretion in handling embassy affairs in Macao. The Jesuit connection, in turn, may have drawn the embassy even more into Macao factionalism, which often had a friars-versus-Jesuits aspect. The connection was strengthened in October 1668 when the Jesuits sent Saldanha, at his request, Francisco Pimentel, S.J., to serve as his chaplain. The Jesuits also provided a learned Christian Chinese to serve as Chinese secretary to the embassy. By the time the embassy left Canton, Saldanha was complaining of the "diabolical spirits of Macao," while Bento Pereira de Faria was accusing other members of the embassy party of intrigue and illicit possession of the embassy's travel documents.[82]

Some kind of news of a new message from Peking approving the embassy was known in Macao by December 10. On December 30, the presents for the Emperor, or possibly just a list of them, finally were shown to the high officials, and presumably were reported by them to Peking. But the presents were embarrassingly mean, "the benzoin a few miserable grains, the amber like gravel, the coral very small . . . the rose water is spring water."[83] Saldanha hoped to be able to improve the quantity and quality of these presents, but, with the dispatch of a list of them to Peking by the Canton officials, the way should have been cleared for the formal acceptance of the embassy.

The arrival of this report in Peking, probably late in January 1669, came at a most opportune time, following upon a turn of fortune worthy in substance and timing of a baroque religious playwright. On 25 December 1668, the three Peking Jesuits, rising to celebrate their fourth Christmas in enforced isolation and idleness, were visited by messengers from the Emperor who at once began to question Verbiest on the errors in the calendar for 1669 compiled by Yang Kuang-hsien and his anti-Christian associates. The next day, the two sides confronted each other before the young Emperor, whose poise and mastery of the situation made a great impression on Gabriel de Magalhaens, S.J., and Yang's astronomical incompetence was exposed. After a series of dramatic tests of the accuracy of their astronomical observations, at the end of February 1669 Verbiest was named head of the Astronomical Bureau and Yang was placed under arrest.[84]

It seems likely that the Emperor was inclined to favor or at least tolerate the missionaries not only for the sake of their knowledge of the heavens but also because of family connections. He must have heard stories of his father's friendship with Schall; his mother was a member of the very unusual and very important T'ung family; and one of his empresses was a daughter of one of his T'ung uncles. This family, shifting back and forth between Manchu and Chinese ethnic identifications, was deeply involved in the politics of the Manchu Imperial House from the rise of Nurhachi to the Yung-cheng succession crisis. Two of the Emperor's T'ung uncles reportedly were sympathetic to Christianity, and a third, T'ung Kuo-ch'i, was married to the "Lady Agatha" of the missionary accounts, paid for the rebuilding of the church in Foochow, wrote prefaces to several missionary works, and was baptized in 1674.[85] Another pro-Christian influence on the Emperor was Songgotu, the dominant high official of the early years of K'ang-hsi's personal rule, who was viewed by the Peking Jesuits as a sympathizer and protector.[86] There are hints in our missionary sources that, especially in managing the changes in Canton after the fall of Lu Hsing-tsu, officials sympathetic to Christianity may have regained influence in the bureaucracy and/or found it prudent to express their sympathies more openly even before December 1668. We have seen that one of the high officials sent from the capital was an old friend of Schall and was ostentatiously friendly to the Jesuits in Macao, and that this visit probably gave the Shang family courage to resume personal contacts with missionaries. Even Chou Yu-te, the new Governor-General, had a Christian connection; in March 1668, the Peking Jesuits wrote to their colleagues in Canton, "by way of the mother of Viceroy Chou." The Peking Jesuits also would do all they could for the embassy; already in April 1667 they had written to a colleague in Macao that an embassy would be very advantageous both to the missions and to Macao.[87]

The Peking Jesuits recorded that (presumably in reply to the report of the showing of the presents on December 30) in the first month of the new year (1 February–1 March 1669) orders were sent that the embassy was to leave for Peking at once; later they learned that the Emperor was expecting it to arrive in Peking in May. In their letter to their Canton brethren on the February decision, they wrote that, if they had known of this in time, they would have urged that the Ambassador's departure be delayed until the eleventh month (November 23–December 22) of 1669 so that he would not be in the north in the winter and have to spend a lot of money on furs for himself and his suite. I suspect that the Peking Jesuits discussed this suggestion with their friends in the bureaucracy, and that it contributed to a decision in September or October that the Ambassador's departure be delayed until the eleventh or twelfth month.[88]

When the February order was received in Canton early in March, it was understood that the Ambassador was to show his credentials to the officials, and, if they were in order, the embassy was to start north at once. Saldanha tried to insist on going to his meeting with the officials in a splendid sedan chair he had had made at great expense, which required 8 bearers, but the officials said his chair could have only 4. Then he learned to his great indignation that the party going to Peking could include only himself, 3 principal subordinates, and 22 others, and that even slaves would be counted in that number. Finally he had to accept this decision, and there were only 13 Portuguese in the party that went to Peking.[89] But months passed and he did not leave. Probably Saldanha still was protesting the limitations on the size of his suite and hoping for reconsideration. Probably he thought he would be able to improve the presents after Macao's ships came in the summer of 1669. Almost certainly he was ill much of the time.[90] In June, Saldanha wrote that he had been detained for two months "solely by lack of money," and that another order already had come for him to go to Peking.[91] Had no one bothered to tell him that boats, coolies, cash subsidies, and at least some food supplies would be provided at imperial expense once he set out? Even more striking is the Canton officials' toleration of these delays, especially in view of their Foochow counterparts' anxiety to send Van Hoorn on his way early in 1667. Probably they were not at all unhappy to have the embassy still in Canton as a cover for further trade with the Portuguese. And, unlike the Foochow officials who may already have been in disfavor for reasons unrelated to the embassy, Shang K'o-hsi and Chou Yu-te were parts of the new and rising alignment linked to the Emperor's personal rule. Most important, I suspect, to the degree that the Canton officials were anxious to promote trade with the Portuguese and a rapprochement with the missionaries, they must have been most unwilling to send on to Peking, where the political situation still was uncertain and the enemies of Catholicism still were powerful, a Portuguese Ambassador whom they had found tactless, incompetent, and neither able nor willing to give presents and exchange courtesies.

The Emperor became more cordial to the Jesuits in April and May of 1669, summoning them for long conversations about astronomy and about Europe, listening to their statements on behalf of Macao. The situation took another dramatic turn on June 14–16, with the Emperor's imprisonment of Oboi and his followers. The Emperor summoned the Jesuits as soon as the crisis began, and later sent a eunuch to ask them if they were pleased by the arrest of their adversary. In the following weeks, he had a number of conversations with them, asked them questions about Europe and about mathematics, and sent them fish from his lake. When the Jesuits sought to inform him of how Lu Hsing-tsu had tried to extort 240,000 taels from Macao he said, "It wasn't 240,000, but 280,000." He had read the file from his inspectors, and remembered. Thus, from the very day when he took all power to himself, this magnificent fifteen-year-old was showing special favor to the Jesuits, not only, I think, because he was fascinated by their distant lands and their scientific knowledge, but also because that favor so dramatically distinguished his regime from that of his Regents.[92]

This situation, of course, was much more propitious for a Portuguese embassy, whatever its failings. On September 1, Saldanha wrote that yet another order for his departure had come.[93] Late in October, the Emperor summoned Verbiest and said that he could not understand why the embassy had been delayed so long, but now a Board of Ceremonies official had told him it would not come until spring, because the rivers and canals would be frozen in the north in the winter. The Peking Fathers remarked that this would free the Ambassador from the great expense of buying furs. The Jesuits continued to feed the Emperor's interest in the embassy, compiling and presenting to him a book giving a very favorable account of the King of Portugal, the extent of his conquests, the organization of the government, and the nobility and ancient lineage of the Ambassador. The Emperor asked them many questions about Europe, and even had them examine one of the Persian horses that had been brought by Van Hoorn over two years before.[94]

FAVOR AND A FUNERAL

The Ambassador and his party embarked on their boats on 4 January 1670, and made final preparations for their journey.[95] As they passed through the streets on their way to the boats they made quite a show in their splendid new clothes. Saldanha's were of crimson satin with silver trimmings. A huge crowd turned out to see the spectacle, and some of the high officials had the windows of their mansions opened so that they could look on. On the lead boat Saldanha's room was richly decorated with carpets, furniture upholstered in velvet, damask curtains, and a damask canopy with gold fringe above the royal letter and portrait. The Ambassador's boat also bore a Portuguese royal standard[96] and a yellow banner with characters identifying the occupant as the "Ambassador from the Great Western Ocean coming to present congratulations" (*chin-ho*). The Jesuits somehow had arranged this in place of the more usual "to present tribute" (*chin-kung*). They insisted, and many later Jesuit and Portuguese sources repeated, that this represented an epochal victory over the megalomania of the tribute system, that for the first time an Ambassador was being announced and received by the Emperor as an equal.[97] Actually, there were ample precedents in the Ming tribute regulations for the use of *chin-ho* for the first tribute embassy sent from a particular sovereign to congratulate a recently enthroned emperor.[98] But I doubt that the Jesuits knew this; their deception of the Portuguese on this point is more likely to have been completely innocent than their deception of the Ch'ing officials about the authenticity of the "letter from the King of Portugal."

All European embassies to Peking were occasions for Europeans to satisfy their curiosity about the fabled city and were meetings between two cultures, two mentalities, that went far beyond the factors involved in negotiations and embassy procedures. For most embassies, the sources are so matter-of-fact or so unsatisfactory that it is hard to sense these broader dimensions. For this embassy, however, we have the rare treasure of an eyewitness account by Father Pimentel, the Ambassador's chaplain. Yangtze River sturgeon, pious Chinese Christians, Peking bedbugs, a sheep's head served at a banquet, the fatness of Manchu officials and the meanness of the streets and buildings of their capital, all figure in it along with excellent information on the embassy; it is translated as Appendix A to this book. The rest of this chapter is based largely on it; only points taken from other sources or out of chronological order in Pimentel are noted separately. But these summary pages are no substitute for reading the Father's observations and reactions in his own words.

The embassy left Canton on January 10, heading up the North River, over the Mei-ling Pass, and down the Kan River to the Yangtze. It took almost six months to reach Peking; in 1656 the De Goyer-De Keyser embassy had taken only four.[99] Its pace may have been slowed by Saldanha's chronic illness, which was especially severe in March and April.[100] Also, Pimentel, traveling in secular dress because missionaries still were barred from the interior, stopped at many places to bring the sacraments to congregations that had not seen a priest in five years. North of the Yangtze, low water in the canals may have impeded progress, as it had for Van Hoorn in 1667. In Nanking in Holy Week, Pimentel was summoned to the mansion of T'ung Kuo-ch'i, the Emperor's uncle, was received with great courtesy, said two masses, baptized two children, and heard confessions most of the night.[101] Pimentel noted as special honors several practices that almost certainly were normal for tribute embassies; the payments of silver for expenses, the providing of coolies at regular intervals to pull the boats, the presents—probably mostly food—from many local officials, the exemption from internal customs inspections and tolls. At Huai-an and Lin-ch'ing, the high officials sent subordinates ahead to clear the way for the embassy on the crowded Grand Canal.

Pimentel also recorded with relish two incidents in which the Ambassador and his party insisted on their precedence over Chinese officials and seemed to get away with it. When they crossed into Pei-Chihli, a Chinese official attached to the embassy party, presumably the "accompanying official" or one of his subordinates, went to the local magistrate to seek the usual money subsidy and fresh coolies to pull the boats. The embassy official and the local magistrate got into a heated argument, came to blows, and the local official was wounded, arrested by the embassy official, and brought to the boats where he spent the night trussed up on the deck in sight of his own people, who did not dare come to his aid. The next morning, Saldanha had him released and presented with some European goods. To the Portuguese this incident indicated the extraordinary prestige of the embassy; we may wonder if the embassy official had not pressed his luck a little and risked a report to Peking, but we also should remember that we simply do not know the detailed regulations that governed local officials' behavior toward a tribute embassy, nor what informal sources of power the embassy officials may have been drawing on. Farther north in Pei-Chihli, the embassy met a boat carrying a brother of the Feudatory Prince of Fukien, either the aged Keng Chi-mao or his son Ching-chung who would succeed him in 1671.[102] Apparently, for such a meeting on the Canal, the party of lower rank was supposed to let its towrope go slack and move to the side of the Canal away from the towpath while the superior passed. Keng refused to lower his towrope, and both parties came to a halt, while Keng threatened to refer the dispute to Peking. Then a sudden storm came up, the embassy junks broke loose from their moorings and came to rest at the far side of the Canal, and the next day both parties went their way in peace. Pimentel saw this as Heaven's way of settling the dispute, and Keng probably would have agreed, for, if I read Pimentel's account correctly, it was the embassy's junks that had been blown to the far side of the Canal, yielding precedence to Keng's!

The embassy arrived at T'ung-chou outside Peking on June 27. The three Peking Jesuits, Luigi Buglio, Gabriel de Magalhaens, and Ferdinand Verbiest, came there to greet them, the first European visitors the Fathers had seen in five years, and instructed the Ambassador on the questions Board of Ceremonies officials probably would ask.[103] Accompanied from T'ung-chou by a Board of Ceremonies official, the embassy entered Peking on June 30. It was lodged not in the usual Residence for Tributary Envoys but in the recently confiscated mansion of one of the four Regents. I think this must have been Oboi's mansion;[104] if it was, in lodging an embassy so closely linked to the once-persecuted Jesuits in the former mansion of the supreme Regent, the young Emperor had found the perfect gesture to dramatize his own break with the politics and policies of the Regency. The Peking Jesuits, usually eager to record and expand on any sign of favor, did not make much of this one, perhaps because they had regarded Suksaha, not Oboi, as the arch-persecutor among the Regents.[105]

The royal letter and presents to the Emperor were turned in on July 1. Among the presents, the only one that made a very good impression was a portrait of King Affonso; according to one Chinese source, it had been sent because the distance was so great that the King could not come in person to pay homage to the throne.[106] It was said that one official had complained that the pieces of coral were far smaller than those brought by the Dutch; less plausibly, the same report described how he had been rebuked by his son for thinking more about the value of the presents than about the honor of the sovereign who sent them.[107] Two pieces of cotton cloth were rejected because nude human figures were represented on them.[108]

After the presents and letters were turned in, the Board of Ceremonies officials had several questions to ask the Ambassador. They wanted to know why the King of Portugal had not used the word *ch'en*, "Minister," to refer to himself in his letter. (The reference of course was to the Jesuit translation of the letter, and this suggests that the Jesuits probably had made a fairly close translation, not a free adaptation to "tribute memorial" conventions.) Saldanha, carefully coached in advance by the Jesuits, replied that in Europe it was not the custom for some kings to call themselves vassals of others when writing to them, and the officials apparently let the matter drop. I have found no record of any Chinese version of this letter; I think it probable that another more acceptable "translation" was done for the imperial archives. The Portuguese and the Jesuits were convinced that this was another epochal victory over the pretensions of the tribute system, like that of the banners on the embassy boats. I think it likely that only a few Board of Ceremonies officials ever were aware of the problem, but, if they were willing to gloss over the breach of tribute protocol in the royal letter, it probably was because the Emperor was taking such an ostentatious personal interest in the embassy. To raise procedural questions about it would have been to risk the displeasure of this brilliant and willful adolescent.[109]

Saldanha also was asked if the King of Portugal would send another embassy, and replied that he had no way of knowing. Probably he was asked many more questions, especially about his country, its government, and its products; Pimentel simply records that they returned to their lodgings "giving a thousand thanks to God because he had delivered us with honor from such ignorant questioners." Thereafter, Saldanha's illness became much more grave, and he remained in bed until the end of July. The Peking Jesuits were not allowed to visit him until the Emperor returned to Peking from a hunting expedition.[110] Then the Emperor not only ordered that imperial physicians be sent to treat him, which was a regular part of the procedures for a tribute embassy, but also ordered that the Jesuits visit the Ambassador twice a day and report to the Emperor each time on his health. The weather was very warm, and he had snow or ice sent to the embassy party every day, as well as deer he had killed on the hunt, fish he had caught, and fruits from his gardens. The Portuguese and the Jesuits also understood that the embassy's daily food rations were four or five times more generous than those that had been given other embassies. But the lists they give, which check fairly closely with the lists in Chinese sources, do not show any such discrepancies; at most, it seems probable that Pieter van Hoorn, his son Joan, and Constantijn Nobel had had to share among themselves two top-grade allotments, since embassy regulations anticipated only one primary and one secondary ambassador, while Saldanha had one top-grade allotment all to himself.[111]

On July 30, the Ambassador and his suite went to the Board of Ceremonies to practice the ceremonies they would perform at the audience. They were told they must be very careful to keep their hats on when they prostrated themselves, and that they would not be allowed to wear their swords in the presence of the Emperor. They understood that this prohibition had been introduced after the arrest of Oboi for plotting to assassinate the Emperor during an audience. Saldanha protested this prohibition so vehemently that even some of the Portuguese thought he was overdoing it. The Board officials could not back down on a prohibition that applied to all Ch'ing officials and nobles without exception, but they were so anxious to mollify the favored Ambassador that they made a show of consulting among themselves on the matter three times.

Saldanha and his suite had their routine formal audience on July 31, which was an "ordinary audience" day.[112] When the officials came to conduct him to the Palace, they told him he would have to have the cover removed from his sedan chair, because it was of crimson and gold cloth, colors reserved for the Emperor. And, of course, no one except the Emperor and a few highly honored senior officials rode chair or horse within the Palace precincts. Saldanha, who had spent the previous month sick in bed, and still looked more dead than alive, had to walk about 800 meters from the T'ien-an Gate to the great courtyard before the T'ai-ho Hall. After the basic homage ceremony was completed, the embassy party was escorted into the side of the Hall, as Van Hoorn's had been. But, while Van Hoorn had waited in vain for the Emperor to approach him, he did come to Saldanha and speak to him very affably. As the Portuguese were returning to their lodgings, they were summoned again to an informal interview at the Ch'ien-ch'ing Gate. This gate, behind the audience halls, is a large open structure—Pimentel rather aptly called it a "veranda"—which frequently was used for the Emperor's working audiences with his high ministers. This summons meant another and even longer walk for poor Saldanha. Fathers Verbiest and Buglio also were summoned, and Verbiest interpreted. Unfortunately, no record of what was said seems to have been preserved. The Ambassador and his suite received at this audience gifts of damasks, velvets, and other fine silks. Their bestowal at an informal audience with the Emperor, the simple fact of such an audience at the farthest point into the palaces where any high official would be received, the imperial summons to the Jesuits to participate, combined to make a very impressive demonstration of imperial favor both to the embassy and to the Fathers.[113] On August 14, the Emperor ordered that, since this was the first time the Portuguese had come to pay tribute and since they had come from such a great distance, their presents should be especially generous; this probably is a retroactive sanction for the presents given on August 31. The August 31 presents included nothing for the King of Portugal, and were not included in the totals of gifts recorded in the Ch'ing compendia of regulations, but probably provided a precedent for the more generous gifts to the Pereira de Faria embassy.[114]

About the middle of August, Saldanha and his suite were summoned to receive their more routine imperial gifts; Pimentel noted that, since they were not entering the presence of the Emperor, they were allowed to wear their swords. The usual three banquets followed. Pimentel was emphatic about the

Sources: Dapper, *Gedenkwaerdig Bedryf*, pp. 364–365 and plate between these two pages. For Chinese charts see *Ta-Ch'ing hui-tien t'u*, pp. 1531–1532; *LPTL*, charts of banquets, 5.

horrors of Manchu cuisine; he was served the head of a sheep at each of the first two banquets.

Pimentel and other Jesuit sources suggest that the embassy was treated well because of the Jesuits' repeated emphasis in conversations with the Emperor and writings presented to him on the prestige and high rank of the Ambassador. He was received as a count, was referred to popularly as a petty king, and the Emperor did not refer to him as a minister of the King of Portugal. The Peking Jesuits also made sure of the support of two very powerful and reasonably friendly officials by giving them presents in the Ambassador's name: a fine clock for Mingju and gifts sent by the Macao and Canton Jesuits for Songgotu.[115] These suggestions may not have been totally without foundation, but the most important cause of the Emperor's emphatic gestures of favor to the embassy almost certainly was his own desire to dramatize the great transition from the rule of the Regents to his own personal rule. In addition, he probably had a personal interest that went beyond the political in the Ambassador and his exotic suite. On August 9, he again sent officials to inquire about Saldanha's health.[116] Some time after the banquets, the whole suite again was summoned to the Ch'ien-ch'ing Gate. Pimentel's account of this interview is to me one of the most interesting and amusing of the many accounts of K'ang-hsi by European eyewitnesses; the young Emperor, seated in the great open gate, discussing falcons, hunting dogs, weapons (Pimentel thought he was hinting these were presents he would like to receive), taking a good look at an African slave, probably the first African he has ever seen, checking to see if the color can be washed off, while his palace ladies peer through the cracks in the great doors at the back. Also, of course, the summons to a second interview there was an altogether non-routine and signal honor.

All these signs of imperial favor probably did a great deal to help assure the survival of Macao, but did nothing directly to resolve its particular difficulties with the Kwangtung officials. Indeed, our sources tell us little about how the Macao authorities would have articulated their grievances or what specific changes, beyond some form of legalization of their maritime trade, they would have asked for at this time. In Canton the Jesuits had already advised Saldanha to deny that he had any specific requests to make or business to conduct on behalf of Macao, and, up to this time, he had followed that advice. But, before the suite entered the palace for its second informal audience, Bento Pereira de Faria, secretary of the embassy and one of Macao's representatives in the suite, handed Saldanha a memorial describing the services of Macao to China in campaigns against pirates and in resistance to the Dutch, and requesting some form of legalization of the City's maritime trade. Saldanha at once asked the Jesuits if it would be wise to present such a memorial; they strongly advised against it. The references to Portuguese deeds of arms would be most unwise, they said, since the Ch'ing already had shown in their relations with the Dutch how wary they were of foreign sea power near their coasts; it was most unlikely that maritime trade would be opened while the Cheng regime remained on Taiwan, and a request now would produce nothing but a negative precedent that would be hard to overcome later; and, above all, the Ambassador had denied from the beginning having any Macao business to discuss, had been esteemed more because he had not come to engage in trade or ask for anything, and would lose respect and credibility if he now presented such requests. Saldanha did not present the memorial, but he did bring up the plight of Macao in conversation with the Emperor, who replied "that he already knew all that." According to Pimentel, the Jesuits had been informing the Emperor of the plight of Macao in their conversations, and had even given him "a printed book" on the subject. From these and from his official sources he no doubt did know a great deal, and his personal interest in Macao may have helped protect it against a return of the gross extortions of the 1660s. The experience of the Dutch suggests that the tribute embassy was not a good medium for substantive negotiation and the making of binding commitments. The Jesuit argument against asking for legalization of Macao's maritime trade at that time was a plausible one. But, to Bento Pereira de Faria and many people in Macao, it seemed that the Jesuits had simply betrayed Macao and sabotaged efforts to negotiate on its behalf in Peking.

The Emperor was reported to have suggested that the embassy not leave Peking until early October, so that Saldanha could have more time to recover his health; he may also have thought that the fall rains would make the Canal trip quicker and easier. But Saldanha replied that he had to leave by the end of August in order to arrive at Macao in time to sail for Goa in the north monsoon, which would end in March or April, and the Jesuits urged immediate departure in conversations with the Emperor.[117] On August 20, he was permitted to spend the day visiting the Peking Jesuits. On August 21, he was summoned to receive the imperial edict to the King of Portugal, and left Peking the same afternoon, reportedly now permitted to use the crimson and gold cover of his sedan chair, surrounded by large curious crowds. Shortly after the party left Peking, two African slaves escaped from it, but, when they were caught, officials were sent to return them to the embassy, with travel documents to bring them all the way to Macao if necessary. Saldanha's associates thought his health was much better, but, on October 18, on the Grand Canal north of the Yangtze, he fell ill again, probably with a stroke, never regained consciousness, and died in Huai-an on October 21.

Saldanha's death was reported to Peking at once. Verbiest reported that the Emperor expressed his grief "immoderately." The mourning ceremonies ordered by the court are described by Pimentel; they seem to have been those prescribed by the Ch'ing regulations for the death of any tribute ambassador.[118] When the orders caught up with the embassy in Canton, the hsien magistrate kowtowed and burned incense before the coffin, made offerings of food and wine, and read an imperial mourning proclamation, and Prince Shang K'o-hsi and the provincial officials also sent officials to pay homage.

After Saldanha's death, Bento Pereira de Faria took charge of the embassy. Our only source on this change is by the Jesuit Pimentel, bitterly hostile to Pereira de Faria and written to refute his calumnies of the Jesuits. He makes it a black farce: Pereira de Faria raised a ruckus during the Ambassador's death throes, got the soldiers on his side by raising their pay, rejected the claim of the man whom Saldanha had appointed his successor, and had his own boat put in the front of the procession, leaving Saldanha's body in a cheap and leaky coffin on the second boat, to the amusement or indignation of the accompanying Chinese. Pimentel says Pereira de Faria even disregarded express orders from the Captain-General of Macao to yield precedence to Saldanha's boat for the embassy's return to Macao, which took place during Holy Week (March 22-29) of 1671, almost four years after Saldanha's departure from Goa.

This was not quite the end of the black farce. Bento Pereira de Faria printed in Macao a manifesto accusing the Jesuits of sabotaging the embassy; Luis da Gama, S.J., replied with a prolix and legalistic "Just Defense," drawing on Pimentel's account and asserting that Pereira de Faria's libels ought to be punished by excommunication.[119] Saldanha's widow, probably in Portugal or in Goa, wrote to the Prince Regent to accuse Alvaro da Silva, Captain-General of Macao, and an associate of having caused her husband's death "in the City of Macao" when he had been "sent as ambassador to Japan."[120] In April 1672, Bento Pereira de Faria arrived in Goa bearing the Emperor's letter, a copy of which was sent on to the Prince Regent, and the presents, which were turned over to representatives of the Princess. The Viceroy reported to the Prince Regent that the Macao authorities had informed him that nothing good had come of the embassy, "for the tribunals of Canton, which were very fearful that some reprimand or punishment would come to them along with the Ambassador when he returned from Peking were disabused of this, and nothing of the kind having come, and seeing themselves free of this care, they again molested and vexed the said City, with repeated chops to take silver from it."[121]

The extortions did continue, as will be shown in the next chapter. No document had been issued in Peking or in Canton as a result of the embassy that in any way changed Macao's status. Indeed, the known Chinese documentation on this embassy is remarkably slim; the only mention I have found beyond routine bureaucratic sources and a few private reflections of them is the brief note on the King's portrait mentioned earlier. The bureaucratic sources mention Saldanha's death and the official mourning for him. They also indicate that no fixed tribute period was established for Portugal because it was so far away; perhaps the first faint sign of acknowledgment of the irrelevance of such bureaucratic categories to the management of relations with Europeans.[122]

So the Viceroy and the Macao authorities might well have asked how they would have been worse off if they had avoided all that expense and not sent an embassy at all. Those officials who argued for the extinction of Macao then would have been able to argue that Portugal had no standing at all in the Chinese world order, but it is not clear how strong an argument this would have been. Ming and Ch'ing rulers had accepted Macao even though Portugal had never sent an embassy; and the last mention we have of a proposal of extinction is from 1670, after the acceptance of the embassy. Saldanha had not tried to use the embassy as a channel for substantial negotiation, but I see no reason to think that he would have accomplished much if he had. Negotiation was better left to the Jesuits, who were at court all the time and had a growing range of informal connections and friendships. The arrival of the embassy had given the Emperor an excellent opportunity to show favor to both the Portuguese and the Jesuits. The Jesuits were rising in his favor for reasons that had nothing to do with Portugal or Macao, and their influence and prestige would have helped to protect Macao even if no embassy had come. But those gestures of favor to the embassy probably made the link between Macao and the Jesuits clearer to many high officials than it would have been if no embassy had come. I suspect that the Macao city fathers would have said that that was not much of a return on an investment of over 30,000 taels.

chapter four

Bento Pereira de Faria

1678

This very short chapter draws together all the information I have been able to find, and probably very nearly all there is, on the least well-prepared of our four embassies, the one with the most thoroughly fradulent documentation, and the one that accomplished the most. The paradoxes of our subject are thrown sharply into relief. Planning, expenditure, and diplomatic competence counted for very little. Spokesmen with permanent residence and good relations at court were more important than all the above put together. And the great shifts of Ch'ing domestic politics, fundamentally without connection to foreign relations, were more important still. No present made as great an impression as an exotic animal, but, without the right internal political conditions, we may doubt if a whole African menagerie would have accomplished much.[1]

MACAO: PRECARIOUS SURVIVAL

Macao had been saved from destruction in the summer of 1667, but the prohibitions of its maritime trade remained in force. The 5 ships that arrived in 1667 apparently all were reported as having arrived as part of the Saldanha embassy, and thus exempt from the ban and admissible for trade. Four that arrived in 1668 were allowed to trade, but had to hide in outer anchorages when officials came to inspect the coastal areas in connection with the revocation of coastal evacuation. Eleven Portuguese ships arrived in the summer of 1669, including some that Macaenses had bought or had built in other ports; Macao's trade probably remained at or near this level in the following years, but we have only fragmentary information on it. At the end of 1668, there were hints of another attempt by Ch'ing officials to extort a huge sum—120,000 taels—out of Macao in return for legalization of its maritime trade, but, to judge by our only extensive source on this period, this attempt evaporated after a few weeks.[2]

Beginning in 1669, Macao faced new difficulties resulting from overseas Chinese and Dutch "free burgher" residents of Batavia trading in the islands near Macao. This was unwelcome competition for Macao's trade between China and Southeast Asia and also, since all maritime trade was prohibited by the Ch'ing Government, led to renewed difficulties between the city and Ch'ing officials who accused the Macaenses of tolerating the presence of the ships and engaging in trade with them. One of the Loyal Senate's many accusations against Alvaro da Silva, Captain-General from 1667 to 1670, was that he had tolerated the Dutch burgher presence in the islands, and Macaense trade with them, and encouraged them to come again.[3] His successor, João Borges da Silva, also tolerated the Dutch trade, but did try to do something about it; he sent an envoy to Batavia with a letter complaining that the presence of the burghers was threatening the small beginning of improvement in Macao's relations with the Canton authorities, and asking that Dutch ships be forbidden to go to the Macao area. The envoy also presented a secret proposal for a joint Dutch-Portuguese venture in trade between Macao and Batavia, but both this and the request that voyages to Macao be prohibited were turned down.[4] The complaints were renewed in letters from Borges da Silva in 1672 and from his successor, Antonio Barbosa Lobo, in 1673, but the requests to stop Dutch burgher trade were not repeated.[5] There were 10 Dutch and overseas Chinese ships trading near Macao in 1672, 12 in 1673, 11 in 1674.[6] One source says the officials demanded 17,000 taels for permission to trade in 1671, and one military official demanded 11,000 for himself alone in 1672 and raised it to 15,000 in 1673, but it is not at all clear how much was paid or to whom. Other sources suggest that the City was exploring through the Peking Jesuits the possibility of legalization of its maritime trade in return for a payment of

50,000 taels.[7] We know almost nothing of the internal politics of Macao in these years. Antonio Barbosa Lobo, Captain-General from 1672 to 1676, at first was found satisfactory by the Senate, but, in 1676, the Dutch reported that there was much resentment against him, and it was possible that he had been sent away in chains. Even less is known of his successor, Antonio de Castro Sande, 1676-1679.[8]

Perhaps more important for the city's survival were the dramatic events in south China in these years. In April 1673, Shang K'o-hsi requested permission to retire and to transmit his princedom to his son, Shang Chih-hsin. The Emperor allowed him to retire but denied the request for succession and abolished the princedom. The other Feudatory Princes, Wu San-Kuei and Keng Ching-chung, then had to make the gesture of requesting the abolition of their princedoms as well; K'ang-hsi, overruling most of his senior advisors, accepted the requests, and Wu rebelled, beginning the "Rebellion of the Three Feudatories" that for a time seemed about to overthrow the Ch'ing dynasty. In Canton, Shang Chih-hsin was generally in control of military affairs, but his father still had considerable influence; they finally allied with Wu San-kuei only in March 1676.[9] From 1673 on, they had little reason to fear surveillance from Peking and were anxious to use maritime trade to build up their own wealth and power. In 1674, for example, Shang Chih-hsin wrote to the Dutch Company authorities in Batavia to encourage them to resume their trade to Canton.[10] The Shang family had been on good terms with Portuguese and missionaries for many years, and these years probably were an interval of relative relaxation in Macao's relations with Canton. Dutch records of trade in the Canton area late in 1676 show a great deal of political tension and instability in Canton but give no indication of Macao being in any particular difficulties. The Dutch also reported that the Portuguese told them they had fought a pitched battle against Cheng forces in the area after a Cheng fleet tried to keep Portuguese ships from leaving until the city paid over 8,000 taels.[11]

The Fukien and Kwangtung components of the Rebellion of the Three Feudatories collapsed in October 1676 and January 1677. The Ch'ing Government allowed Shang Chih-hsin to retain a large measure of power in Kwangtung, and began to press him to contribute to the campaigns against the remaining Wu forces in the southwest and Cheng forces on the Fukien coast. By 1678, the inland river route was clear for an embassy to go north, and the time was peculiarly propitious. The Peking Jesuits, closely linked to the Portuguese and to Macao, had continued to rise in the Emperor's favor, as they demonstrated the superiority of their astronomy, taught the Emperor Western astronomy, mathematics, and music, and cast (and blessed!) cannon for his forces to use against the rebellious feudatories.[12] The Emperor himself was emerging as the triumphant consolidator of a new order. The great gamble of his opposition to the feudatories was paying off, and he was devoting more and more attention to presenting himself as a patron of Chinese statecraft, arts, and letters. What better time to send an exotic animal in the "Hounds of Lü" tradition, to be commemorated by court poets and to fascinate the Emperor, hunter of bears and tigers, seeker of knowledge, ruler and sometimes predator?[13] If all went well, it might even lead to some concrete advantage for Macao.

A LION FOR THE EMPEROR

When Bento Pereira de Faria and other prominent citizens of Macao came to Goa in April 1672 the Viceroy received, either directly from them or by letter from Macao, a very interesting suggestion: the Emperor would like to receive a lion. If the Portuguese would send him one, it would do much to facilitate the revival of Macao's commerce. This suggestion may have been based on Jesuit reports of the young Emperor's interest in the hunt and in exotic animals, including the animals brought by the Van Hoorn embassy. The Viceroy responded to this suggestion with unusual promptness, ordering the captain of the fortress of Mozambique to obtain a lion and send it to Goa. By December 1674, it was in Goa and was to be sent to Macao on the next monsoon. There was a year's delay, perhaps connected to difficulties in designating and sending the next Captain-General. The lion was sent in the spring of 1676, and is next heard of on September 13 of that year, when the Dutch recorded that a subordinate of Shang Chih-hsin, who had been out to Lampacao to deal with the Dutch, came to Macao to see the lion, but the Portuguese made some excuse and would not let him see it.[14]

There was a small difficulty about sending the lion to K'ang-hsi; in 1676, the Shang family was in rebellion against him. (If they had remained so longer, might someone have suggested sending the lion to Wu San-kuei?) After Canton returned to Ch'ing allegiance in January 1677, the way north through Kiangsi still was not clear; Ch'ing forces took Chi-an in April, but did not secure all of southern Kiangsi until the end of the year, and the balance of forces in Hunan did not turn against Wu until the spring of 1678.[15]

We do not know when or in what terms the lion first was reported to the court. We do not even know whose idea it was that it should arrive there in the company of what was stated to be a full-fledged embassy from the King of Portugal. Buenaventura Ibañez, a Spanish friar who was in Macao early in 1678, wrote that the arrival of the lion had been reported to the Emperor, he had ordered it sent to him, and Macao now would have to name an ambassador to accompany it. But a Spanish visitor would not have been told everything in Macao, and plans probably were farther along by this time than his information would suggest.[16] Letters from the Goa authorities written before news of the sending of the embassy reached Goa say nothing about an embassy or about the letters it would take; after such information had been received, Goa approved the choice of Bento Pereira de Faria "and the letter which that City wrote to the said Emperor, in the name of His Highness."[17] The eighteenth-century Franciscan historian José de Jesus Maria, not always reliable but apparently drawing on Macao archives that have not survived, is one of our fullest sources on this embassy. He gives the text of one "royal letter" and the gist of another, and seems to have thought that both actually had come from Lisbon. The text he transcribed is notable for its flowery rhetoric and for its frank avowal of Chinese sovereignty over Macao; the latter is another reason for suspecting that it was written in Macao, not in Goa. At several points it is quite close to the text recorded in the Ch'ing *Veritable Records.* The Portuguese text opens with thanks for the Emperor's generous reception of the Saldanha embassy, and says that Bento Pereira de Faria gave a full report of it to the Viceroy in Goa.

"To the same Viceroy of the Indies I ordered that he proclaim in all my Realms and Lordships of the Orient where my subjects live and where my power prevails a firm and perpetual peace between them and the subjects of Your Majesty, and I hope before God that Your Majesty will order it preserved as long as the sun and the moon shall endure . . . I send with this as my envoy the same secretary of the [Saldanha] embassy, Bento Pereira de Faria; I ask that Your Majesty hear and give credit to all that he shall say in compliance with my will, and all that I command him to ask of Your Majesty on my behalf in the way of favor to be shown to my subjects who have lived in Macao, territory of Your Majesty, with all the loyalty and obedience which I have commanded them to show, for more than one hundred years."[18]

The embassy set out for Peking early in 1678, probably before March 5.[19] José de Jesus Maria says that, because the Canton authorities had insisted on seeing a copy of the royal letter brought by Saldanha, this embassy would take with it an unsealed copy in Chinese. He says the Macao authorities "also had received letters from the King which authorized them to propose to the said envoy the matters and transactions which they would judge most appropriate for the common weal and preservation of this City, permitting that in his Royal Name they be presented to the Emperor"; I suspect this "royal letter of authorization" also originated in Macao. The memorandum drawn up under this authorization outlined the following requests to be made in Peking:

1. The Portuguese had lived in Macao for many years, but found themselves so poor that they could not sustain themselves, for they had no farming land, but lived entirely by trade. Permission was to be requested for them to send their ships "to the lands where they were born and had relatives," seeking means to sustain themselves, "coming and going freely without paying measurement dues or any other tolls."

2. The Ambassador was to request "that this City enjoy in this point of land all the privileges enjoyed by all the lands where there are tombs of ambassadors, since the Ambassador Manoel de Saldanha is buried in it." [What could they have had in mind?]

3. When it is necessary to place before His Majesty matters of grave importance for the welfare of Macao, that residents of Macao be able to come to Peking freely, without depending on communication via the Canton officials or permission from them. [Unhindered communication with the court also was sought by the Macartney embassy over a hundred years later. It was gained only at gunpoint in 1860.]

4. That the people of Macao be allowed to go to Canton to trade instead of waiting for Chinese merchants to come to Macao to trade with them.[20]

The embassy party consisted of Bento Pereira de Faria as Ambassador, accompanied by his son and possibly other members of his family; another Macaense, Manoel de Aguiar Pereira, as his second in command and replacement if necessary; three Portuguese whose functions were not specified, a chaplain, a surgeon, a "Papango" (Filipino?) lion-tamer, and six slaves. Almost all the money for expenses was borrowed from Macaenses; profits from Siam voyages that were supposed to be applied to the repayment of the debt to the King of Siam also were held back to make a small addition to these funds. The entire sum spent was about 2,750 taels; even doubling the sum to allow for the expenses of bringing the lion all the way from Mozambique, quite a bargain compared to about 30,000 taels spent on the Saldanha embassy. Some money and goods were taken to be given as presents along the way, but the lion was the only present destined for the Emperor.[21]

We know nothing of the embassy's trip from Macao to Peking. The sole present already had been reported to the capital, and, with the Emperor completely in control and no doubt eager to see the lion, it was most unlikely that anyone would raise sticky questions about the authenticity and credentials of the embassy as the Canton officials had done with Saldanha. The embassy also was arriving just as imperial favor to Verbiest reached a new peak. On August 27, Verbiest reported the completion of a calendar for the next 2,000 years, compiled at the Emperor's command; in October and November, he was promoted and given a new honorary title, the latter only after the Emperor found the Board of Ceremonies' first recommendation inadequate. The embassy reached Peking between September 7 and 17.[22] Our best information about the embassy's stay in Peking is in a letter from Father Verbiest translated in Appendix B of this book. The sequence of events and the number of meetings with the Emperor is not altogether clear from this letter; what is clear is that the Emperor was pleased with the lion, and showed extraordinary personal favor to the Ambassador, his suite, and the Peking Jesuits, all of which must have caused much comment in high court circles.

The Emperor probably went to see the lion before the embassy's first formal audience. Verbiest placed a short paragraph about it before his account of the audience, and this would agree with the sequence of the Van Hoorn embassy. The Emperor also brought his two sons to see the lion, which Verbiest took to be an extraordinary favor. But who would deny a small boy a chance to see a lion? The Princes were Yin-jeng, born 1674, who caused his father so much grief before he was degraded from his position as Heir Apparent in 1708 and again in 1712, and Yin-shih, born 1672, who was accused of having bewitched Yin-jeng and placed under house arrest in 1708.[23] This may have been the first occasion on which the Princes made a semi-public appearance outside the inner apartments of the palace. The Emperor summoned the Jesuits to join him in viewing the lion, conversed with them, and had them stand near him; rare and precious favors indeed, wrote Verbiest. It does not seem that the Ambassador was included in these favors; neither was Van Hoorn when the Emperor viewed the horses and dwarf oxen he had brought. Such a viewing was a personal act of the Emperor, not part of the ceremonial regimen of the embassy. It probably would have been a breach of court ceremonial for the Ambassador to enter the imperial presence before the first formal audience. The viewing probably took place in the Coal Hill area, where, according to Gabriel de Magalhaens, S.J., there were cages for the Emperor's various wild beasts and even artificial holes for rabbits in the hills, or at another menagerie in the Summer Palace area northwest of Peking.[24]

The Ambassador was received in formal tribute audience in connection with an ordinary audience on September 20.[25] The Emperor ordered Verbiest to accompany the Ambassador to the audience and then on to a special banquet in the imperial presence. This would seem to represent a revival of the early Ming practice of banquets for tribute envoys in the imperial presence, which had only a limited place in the Ch'ing ceremonial order as it finally crystallized. Verbiest wrote that this banquet took place at one of the Emperor's "greatest halls"; this description certainly would fit the Pao-ho Hall, which in the fully-developed Ch'ing order was the location of a spectacular banquet for Mongol princes and nobles on the last day of the old year.[26] But the 1690 *Collected Statutes* explicitly state that the banquets for this embassy followed the precedents set for the Saldanha embassy.[27] At this banquet, according to Verbiest, the Emperor sent dishes from his table to the Ambassador and the Jesuits, another mark of high favor.

More honors followed, both for the Ambassador and the Jesuits. The Fathers were given permission to visit the embassy party whenever they wished. For the translation of the "royal letter" Buglio and Verbiest were summoned, and the interpreter who had come with the embassy rejected. Verbiest was present when these translations were presented to the Emperor (and read out loud to him, if I understand Verbiest correctly); the Emperor had Verbiest read the original Portuguese text out loud to him. This was a completely non-routine gesture of favor and, I suspect, of sheer curiosity to hear how the foreign tongue sounded.

Before Saldanha arrived, the Jesuits had prepared a book about him and about the King of Portugal. This time Buglio wrote a book about lions. His *Explanation of Lions* (*Shih-tzu shuo*) brought to Chinese readers some sound information about lions and a great deal of medieval and Renaissance lore, drawing especially heavily on the works of Albertus Magnus and Ulisse Aldrovandi. Lions eat only live meat. They defend their young fiercely. They do not kill those who squat down in front of them. From this the ruler can learn—in good Chinese and

Renaissance-analogical fashion—the virtue of sparing those who submit, a virtue K'ang-hsi was exercising when it suited him in dismembering the Three Feudatories coalition. The story of Androcles and the lion was recounted, along with Aesop's fable of the lion, the fox, and the donkey, some Latin proverbs and metaphors involving lions, and some lore about the medicinal properties of various parts of the lion's body.[28]

This was not the end of the literary activity brought on by the arrival of the lion. We know of four poems that were written about it, all by scholars from Kiangsu and Chekiang; one already was a Hanlin scholar, and two more would pass the special *po-hsueh hung-tz'u* examination early in 1679. One of the latter was Mao Ch'i-ling, already in his fifties, who had served briefly in the court of the Ming loyalist Prince of Lu in 1646. Dense with classical, historical, and literary allusions, the poems celebrate the power and ferocity of the lion and the splendor of the ruler who has attracted such a gift from so far away, and suggest that the present is a golden age of good government and widespreading empire like the Han and T'ang. For the Emperor, seeing the Feudatory threat receding, drawing more and more lower Yangtze literati into his service, growing in his own ability to appreciate Chinese literature, this was heady stuff indeed. As suggested by G. Bertuccioli, these poems may have been written in connection with the festivities around the Mid-Autumn Festival at Wan-shou-shan in the Summer Palace area; this festival fell on September 30 in 1678.[29] That is not the end of references in the literature of these years to a lion supposedly brought by a Portuguese embassy. One other simply repeats some of Buglio's lion lore. Another has a lion taken along on one of K'ang-hsi's southern tours. But we know the Portuguese lion died a few weeks after its arrival in Peking. If there is any substance to the southern-tour story, it may refer to a lion brought in tribute from Central Asia; another source mentions an instance of this in the K'ang-hsi period, and it would not be hard to confuse Portugal (referred to at this time as one of the Western Ocean [Hsi-yang] countries), and Central Asia, the "Western Regions" (Hsi-yü). Other lions had been brought from Central Asia under the Ming. The Emperor's gestures of favor, I think, were responses not so much to the uniqueness of the gift as to the association of the embassy with the Jesuits, whom he had favored so strikingly at the beginning of his rise to personal power, a response heightened by his sense that his personal rule had emerged triumphant from a period when the survival of the dynasty had been in doubt, and by his pleasure with Verbiest's astronomical work.[30]

The Ambassador and his party, accompanied by all four Peking Jesuits, also were taken to see part of the imperial pleasure parks northwest of the city and given another banquet there by imperial order. This was another striking sign of favor, for which there was no provision in the statutes but which was granted to several later European embassies. I think it unlikely that they participated in anything like the mid-autumn gathering of high officials or were in the imperial presence there; Verbiest, emphasizing every sign of imperial honor, mentioned nothing of the kind.[31]

The embassy remained in Peking until November 13.[32] Apparently it received the statutory three banquets, in which the banquets in the imperial presence and at the pleasure park probably were not counted. The presents to the embassy party and to the King of Portugal were distinctly more generous than those recorded in Ch'ing compendia for the Van Hoorn and Saldanha embassies. As noted in Chapter 3, the informal presents to Saldanha and his suite may have provided a precedent for this generosity, but the explicit and formal increase here was a most impressive indication that this peculiar little ad hoc improvisation was being received very cordially.[33] The lion died late in October, and Bento Pereira de Faria wrote to the Macao authorities urging them to keep this news secret, for, if it got out, the Dutch and the French would do all they could to bring another lion.[34] The Emperor, probably drawing on the information in Buglio's book but in any case interested in the medicinal virtues of various meats, sent to ask Verbiest if he thought it would be proper to bury the lion without those parts of his body which were thought to have medicinal properties. Other sources tell us that the lion was given "a sumptuous funeral," and that the Emperor had a marble monument erected on its tomb with an epitaph, "as is done for highly esteemed mandarins." Verbiest does not tell us what he replied to the Emperor's inquiry, and we must leave aside our imaginings of the Emperor sipping cautiously a lion-liver soup or poking with his chopsticks at a lion-testicle dumpling.[35]

It was all very well to have such conspicuous signs of imperial favor, and for such a modest investment. But such signs in connection with the Saldanha embassy had not saved Macao from official squeeze and vexation, and Bento Pereira de Faria had gone to Peking with instructions to make some very specific requests. These requests had been presented, but no reply had been received when the Ambassador left Peking, and further discussions were entrusted to the Peking Jesuits. Some people in Macao thought the embassy had accomplished very little: "The requests were put off a little, since the high officials and the bonzes already had deluded the Emperor." Pereira de Faria, who had been vehement in his condemnation of Jesuit management of the Saldanha embassy, now repeatedly acknowledged the contribution of the Jesuits to the success of his own embassy.[36]

The Jesuits proved themselves capable hands, capable even of getting a clear-cut decision out of the Peking bureaucracy. Eventually the Ambassador's request for formal legalization of Macao's trade with Canton and its maritime trade was granted. From 1681 to 1684, Macao's maritime trade was the only formally legal maritime trade to and from the coast of China, and Macao and its allies occasionally could use this legal advantage to interfere with the trade of formally illegal competitors, such as the Dutch and the English. But the decision did not come quickly or easily, and eventually involved a Peking-Canton political upheaval somewhat reminiscent of the one that saved Macao in 1667.

SURPRISING RESULTS

One of our best sources of the results of this embassy is a letter of thanks from the Macao municipal authorities to the Peking Jesuits.[37] According to it, when the embassy arrived in Peking, the Jesuits had decided that more presents than just the lion would be required to accomplish anything, and Verbiest had given a fine clock worth 600 patacas—very roughly the same number of taels—and by this means secured an order that the Board of War discuss the legalization of Macao's trade. This sounds like a present to the Emperor, but we cannot be sure. Then the Jesuits gave more presents—presumably to officials of the Board of War—to insure that the decision would be favorable, and before this round of negotiations was concluded the Ambassador left Peking, leaving everything in the hands of the Peking Jesuits.

According to José de Jesus Maria, not always reliable or easy to follow but almost our only source on the rest of this story, the next step in Peking's decision-making was the dispatch of inspectors from the capital who came to Macao with orders from the Emperor to find out what kind of place Macao was, what military forces it had, and what number of people. According to a Jesuit letter, this visit took place on 29 January 1679.[38] The inspectors reportedly returned to Peking to report that Macao had no military power, "its people were few and good," and they lived solely by trade. An imperial edict finally was received in Macao in November 1679 ordering that its land trade with Canton be fully legalized at once, and, as soon as the seas were cleared of pirates and rebels (meaning primarily the Cheng forces), the Canton officials should memorialize again to recommend the legalization of its maritime trade. It is not at all clear why this decision did not come earlier. According to a summary by José de Jesus Maria of a lost letter by Verbiest, it had been known for some time in Peking that Shang was impeding Macao's trade in order to favor the trade of the junks he was sending out to ports where the Portuguese normally traded, and later, with the return of the inspectors, it was learned that Shang Chih-hsin had given them 150,000 taels to divide up among the key people in Peking so that no accusations would be made against him. Unfortunately, we have no dates for all this. I think it likely that Shang had persuaded the inspectors to prolong their tour of inspection in the south, or that they had given only a very partial and ambiguous report when they first returned to the capital, and that it was only with the decline of the Songgotu faction, beginning in September, that they decided to make the good report on Macao, which was not in Shang's interest, and that they or others informed the Emperor of the massive bribe they had received. But, by that time, Macao had been forced to contribute almost one sixth of the bribe Shang paid the inspectors, and had been once again on the edge of extinction.

It probably was in December 1678 or January 1679, while the inspectors were still in Canton and Shang was trying to work out some kind of deal with them, that Shang sent four officials to Macao to demand that the city give him 47,000 taels, alleging alternately that this would be a contribution to funds for the war against the Cheng forces and that he had spent 50,000 taels with the inspectors to gain favor for Macao and simply wanted to be reimbursed. The former claim was a plausible one; Shang was in fact being pressed to contribute both troops and funds to various Ch'ing campaigns in 1678.[39] After waiting a few days for an answer to their demands, the officials ordered the Circle Gate closed, all shops shut, all sale of food suspended, and a close siege of Macao by land and sea. The Macao authorities could see no way to raise such a large sum, and recalled earlier cases when the gate had been closed to frighten them but soon opened. They decided to try to temporize by telling the officials they were trying to raise the money, and to hope to await the return of Bento Pereira de Faria from Peking and see how the negotiations with the Emperor had worked out. But, after the food embargo had continued for fifteen days, poor people were starving to death in the streets. At first, there seemed to be nothing to be done except to abandon Macao or to die in futile battle. But finally it was decided that the Diocese, the Holy House of Mercy, and all the churches and religious orders should collect all their silver to turn over to Shang as a pledge of full payment, trusting that God would provide a way to redeem these utensils of the Divine Service from the clutches of the heathen. This receiving general approval, the members of the Senate and others were inspired to find all the treasure they could in their own households, even selling the jewelry of their women to Chinese merchants who suddenly appeared with ready cash. The total assembled came to 23,500 taels, half the amount demanded, apparently usually an acceptable down payment in such cases, plus the worked silver of the churches, which was accepted as a deposit to be forfeited if the rest of the sum was not delivered in August 1680. Shang even had the gall, once he had the unworked silver melted down and assayed, to demand that the City borrow from Chinese merchants to make up a shortage of over 700 taels in the final assay of the 23,500. The Jesuit letter states that over 10,000 of this sum was borrowed from the Chinese merchants.

There the matter rested until the summer of 1680. The church silver was supposed to be forfeited if the second 23,500 taels was not delivered in August. In September, Shang told the Macao authorities he would have the church silver melted down if the payment was not made. According to stories told at the time and passed down in Macao for several generations, Shang went several times to his warehouses to have the silver melted down, but each time was confronted by an old man, thought to be an angel, or by two old men, St. Peter and St. Paul, one of whom threatened him with a sword.[40] Balked in this way, Shang renewed his efforts to obtain the remaining 23,500 taels from Macao. The Macao authorities managed to find ways, very probably through the Peking Jesuits, to inform the Emperor of Shang's extortions. Actually Shang had been on campaign in Kwangsi in the spring of 1680, and had been arrested there. Later he had been accused of trying to procure the assassination of one of his accusers, and, on September 20, the Emperor had sentenced him to immediate execution. Macao's accusations were at best a small piece of the long Ch'ing bill of particulars against Shang, but as in 1667 its place among the victims of the fallen extortioner probably helped to put it in the good graces both of the Emperor and of the new regime in Canton. And, of course, if the Shang regime had not fallen it might have found ways to block full legalization of Macao's maritime trade. Imperial envoys with orders to carry out the death sentence already were busy in Canton in October; they must have been among those who already had brought Shang Chih-hsin captive to Canton and then had sent further accusations against him to the court. From this time on through 1681, agents of Peking were busy dismembering Shang's vast political and commercial empire, confiscating over 400,000 taels for the government and probably far larger sums for themselves and for the Emperor's private funds.[41] In the process, Macao was given back its 23,500 taels, its note for the other half, and its church silver; these may have come directly from Shang's chief client-merchant[42] or from Ch'ing officials who had confiscated them and then thought better of it and returned them to the city.

On 17 October 1680, two officials who were part of, perhaps in charge of, the new order in Canton came to Macao.[43] With their arrival, or possibly even earlier, the Macao authorities received from Verbiest translations of imperial decisions to legalize maritime Macao's maritime trade. No non-Portuguese foreigners were to be allowed in the port, and the customary measurement dues were to be collected. In view of the previous report that Macao's maritime trade would not be legalized until the sea was cleared of pirates and rebels, it must have been decided that this step could be taken now that the Cheng regime had lost its coastal bases and held only Taiwan and the Pescadores. The officials who came to Macao apparently also discussed the legalization of navigation, which still was "promised,"[44] not conclusively granted; probably they had orders to report to the court on the arrangements that would be necessary to regulate Macao's shipping and collect tolls from it. Final confirmation of the legalization was brought by a Canton official some time in 1681, probably early enough to levy tolls on the ships returning on the south monsoon. In 1681, the Dutch managed to trade near Macao despite Portuguese protests that only they were allowed to trade in the area; in 1682, the Portuguese switched to putting pressure on the Canton officials and threatening to inform Peking of illegal trade, and as a result the Dutch and the English got very little trade. In 1683 and 1684, this Portuguese advantage disappeared from view as the Ch'ing conquered Taiwan and moved toward a general opening of maritime trade.[45]

However slight the advantage to the Portuguese was, what is important to us is that it was the result of a clear and definite decision made in response to a request presented by a tribute ambassador. This was by far the most important decision made in response to a request presented by any of our four ambassadors. Ambassadors were in Peking too briefly and were too closely restricted in their movements and communications to accomplish much of this kind. Bento Pereira de Faria did not accomplish it for himself; he left everything in the hands of the Peking Jesuits, who had friends in many high places, knew where and how to give presents, and could take their time in bringing people around to a decision and letting the bureaucracy work through its involved procedures. They were, in effect, a surrogate for the resident embassy which Europeans already discussed occasionally but would not obtain for almost two centuries.

The tribute embassy also served as a stage on which imperial favor for a particular group of foreigners could be displayed, and never more surprisingly and dramatically that in K'ang-hsi's treatment of the Pereira de Faria embassy. But favor to the Portuguese and favor to the Jesuits were linked in every one of K'ang-hsi's gestures, and I think it likely that the favor to the Jesuits was more important to him. Also, it is worth remembering that some of the most important gestures, like the invitation to the imperial pleasure parks, were personal, and had no formal place in the bureaucratic and ceremonial pattern of the embassy.

There are other ways in which the bureaucratic-ceremonial routine seems to slip into the background as we look at this embassy. Already in the decisions on the Saldanha embassy, no fixed periods had been established. Now, in sharp contrast to 1667-1669, there was an extraordinary casualness about documentation, a willingness to accept what the Canton authorities must have known was a sham embassy with sham documents put together in Macao. The Ming heritage of regulation in a "tribute system" was fading. The tribute embassy was taking a more modest place in a Ch'ing pattern of personal imperial direction of many aspects of foreign affairs. In the 1680s, a few more survivals of the Ming maritime tribute system would be sloughed off, and the Dutch would demonstrate that, for those interested solely in trade, the embassy was a completely useless institution.

chapter five

Vincent Paats

1685-1687

In 1686, the Dutch came to Peking for the first time in nineteen years. The embassy routine and its physical setting were little changed, but the Jesuits, under house arrest in 1667, had considerable influence and had introduced the Emperor to a number of facets of European high culture. Maritime trade both by Chinese and by Europeans, illegal and reduced to a trickle in 1667, was fully legal. The English had entered the trade, and French participation also was anticipated.

This account of the Paats embassy embodies an almost too tidy reprise of some of the themes of our study. This Ambassador had several exchanges with the officials concerning his requests, which was more than the other three Ambassadors had managed, but the tribute embassy remained an extremely inadequate channel for negotiation. The influence of the Jesuits and their inner-court connections were very important for the handling of any European embassy; and, if they had their own reasons for being somewhat accommodating to the Dutch, they could hardly be expected to negotiate on their behalf or give them guidance as they had the Portuguese. It seems that the very edict in which the Emperor insisted that the Dutch send an embassy was the edict in which he made the embassy institution irrelevant to the trading powers for a century by opening trade to all foreigners without distinction. Finally, the glimpses of bickering, incompetence, and corruption in this embassy remind us how rare it was that the European powers of this period were able to provide competent personnel for such a venture.

THE SYSTEM SHELVED

Once it had reached institutional maturity in the fifteenth century, the maritime tribute system endured as a concept and a set of bureaucratic precedents for over four hundred years. The precedents for embassy routine were enforced whenever a foreign envoy came to court, but the system in its full elaboration, in which trade was limited to that carried on in connection with tribute embassies, was applied to Europeans only from 1517 to about 1557 and from 1653 to 1685. The Dutch had to send an embassy in 1666 and again in 1685, and each time would not otherwise have been allowed to trade. The Portuguese embassies both were sent to secure for Macao some degree of exemption from the restrictions of maritime trade that seemed to hark back to the pre-1550 Ming system. But the revival of Ming restrictions was in no way a manifestation of Ch'ing principled commitment to that model of maritime foreign relations. It was the restrictive trade policies that produced the apparent revival of the tribute system, not vice versa. These policies, at least occasionally effective against the Cheng regime, forbade all Chinese trade abroad and all non-tributary foreign trade in Chinese ports. But trade in connection with embassies was so strongly sanctioned by Ming precedent (and had been so much a part of the political economy of the Manchu rulers' ancestors) that it was never prohibited or even restricted except by the limits on the numbers of ships an embassy might bring.

From 1680 on, the political imperatives that produced these policies were waning. In March 1680, the Cheng forces, in disarray, abandoned Amoy and Quemoy. The Ch'ing rulers were busy until well into 1681 consolidating their control in Fukien and Kwangtung and appointing the officials who would manage the final assault on Cheng Taiwan and the transition to peace on the south coast. But already they were confident enough to repeal the coastal evacuation orders that had been revived in 1678 and to confirm the legalization of Macao's trade. Further changes came after the Ch'ing conquest of Taiwan in the summer of 1683. The steps taken to legalize maritime trade by Chinese shipping are fairly clear and were completed by October

1684. A policy for handling trade by foreigners in Chinese ports emerged only after a good deal of groping, and some steps in it are very poorly documented.[1] Late in 1683, the officials at Canton suggested that Macao might be made the center for trade with all Europeans, but apparently dropped the idea when the Portuguese opposed it. Then the initiative shifted from Kwangtung to Fukien. When an English ship arrived at Amoy in 1684, a deputy of Shih Lang, the conqueror of Taiwan, declared his suspicion that the arms, gunpowder, and lead in the cargo had been intended for the Cheng regime, but said all would be well if these items were given to the Ch'ing authorities. This was done and was reported to Peking. The English were granted the right to trade free of tolls that year in consideration of their gift, and were told they could come again. Chinese officials sometimes confused the English and the Dutch, calling both "red-hairs," but it is most unlikely that the court was told this was a Dutch gift; at Amoy that same year, the Dutch had to pay tolls. And the English had *no* standing in the rules and precedents of the tribute system.

When a Dutch ship arrived at Amoy later in 1684, the Boards of Ceremonies and War recommended that it not be allowed to trade because it had brought no embassy and a Dutch embassy was long overdue. (An eight-year interval had been established in 1667, and no embassy had come since then.) But the Emperor ordered further consideration of whether it would not be best to open Chinese harbors *to everyone*. This then was proposed and approved. The Dutch were to be allowed to sell the cargo then at Amoy but would not be allowed to trade again until they sent an embassy. We know this key decision only from a Dutch translation.[2] Unless a Chinese text turns up some day to show otherwise, in the light of the permission to the English it seems likely that "to everyone" meant "to all peoples regardless of their standing or lack of standing in the tribute system." The Dutch still had to remove their delinquency within the system, and its precedents would remain applicable if any trading power wanted to send an embassy to Peking; but, for foreign traders and Ch'ing officials dealing with each other in Chinese ports, it had become a thing of the past, to be read about in old Chinese collections of regulations and in European books of voyages.

When the Batavia authorities decided to send an embassy in 1685, they did not have the translation of this key decision or any full account of it. In addition to a second-hand report of it from a Chinese merchant in the Canton area, they were considering reports of a strange encounter on Taiwan in the first months of Ch'ing rule there. Shih Lang had found there a party of Dutchmen and their families, about 18 people in all, who had been held captive by the Cheng regime ever since its conquest of Taiwan in 1661-1662. In making preparations to send the group to Batavia, he treated as its headman one Alexander van 's Gravenbroeck, and had a long discussion with him. It was by no means a foregone conclusion that the Ch'ing Government would want to incorporate Taiwan into their empire, and Shih wanted to know if the Dutch would pay to get it back. Van 's Gravenbroeck thought that unlikely, unless the Dutch also got "free trade" in China and the exclusion from the empire of English, French, Spanish, and Portuguese traders. It is not altogether surprising that he should have been aware of the threat of European competition; the English had traded with the Cheng regime on Taiwan, and there was a good deal of Chinese trade to and an occasional visitor from Manila. He thought the Spaniards already had offered to take over Taiwan, and also had heard, before the end of 1683, that the Ch'ing Government were going to open their ports to all countries. Shih asked Van 's Gravenbroeck if he could make a proposal of the kind they had been discussing on behalf of the Company. He could not, of course, but he did finally submit a statement to be transmitted to the Emperor, expressing gratitude for the release of the captives and mentioning the possibility that the Company might send an embassy the next year to show its gratitude and to request trade and discuss other matters; Shih may have suggested this wording. According to Van 's Gravenbroeck, Shih seemed to think his own power and influence so great that, "in hope of being well rewarded," he could obtain from the Emperor exclusive trading rights for the Dutch, and turn him against all other countries.[3]

This fascinating character, so adept at finding the right patrons and concentrating power in his own hands, here exhibited the over-confidence and braggadoccio the Dutch already had seen in December 1664,[4] and a streak of greed that was disconcertingly frank even by the standards of the time. His confidence in his own future power was especially misplaced at this time; although he had been given all the credit for the conquest of Taiwan and a hereditary title, he already had been summoned to Peking, and, in September 1684, the Emperor remarked that Shih was "a rough warrior who has never studied and has shallow judgment; he is presuming on his merits and becoming overbearing."[5] Shih apparently retained his office of Provincial Commander of Water Forces until his death in 1696, and had considerable power at Amoy when he chose to exercise it, but the Emperor was aware of his limitations and had little need for his talents in maritime warfare.

Van 's Gravenbroeck and his party reached Siam in February 1684, and Batavia in April 1685. From Siam he sent a report to the Batavia authorities, recommending that they send an embassy to Peking via Amoy or Foochow at once. But the Batavia authorities already had reports suggesting that big changes were imminent in Ch'ing policy toward maritime trade and decided to wait another year to see what developed.

The changes in 1684 did, in fact, eliminate some old Dutch grievances. Trade being open to all, foreign merchants would not have to wait for approval from Peking before they began trading. The imposition of tolls on goods and shipping was a new and unwelcome fiscal burden, but it brought with it the dispatch of special imperial toll collectors, whose presence in the ports eventually made it impossible for the high provincial officials to maintain the monopoly of foreign trade in the hands of their client-merchants of which the Dutch had complained so often. But none of this was completely clear from the reports available in Batavia early in 1685, and the political consequences of the dispatch of the toll commissioners only gradually became apparent in Chinese ports.

FOOCHOW AGAIN

In 1685, discussion of China policy was stimulated by the onset of the season for voyages there and probably by the arrival of Alexander van 's Gravenbroeck and his party from Siam early in April. Joan van Hoorn, who had accompanied his father to Peking in 1666–1668 and now was Secretary of the Council of the Indies, took the lead in reviewing reports from China and presenting proposals to the Council. Apparently, most of the discussion took place on May 7 and 8, and on the 8th the Council decided to send an embassy to Peking as soon as possible. They noted that the Dutch trading near Macao in 1684 had been told that the Dutch were obliged to send an embassy every eight years, and that it would be hard for them to get "free trade" until they sent one, but they did not note the quite explicit warning that they would not be allowed to trade at all until they sent an embassy. The embassy was to seek "permanent and peaceful trade" in China and permission to establish one or more permanent trading lodges in Chinese ports.

In the Council's discussions of this embassy and instructions for Paats, the sense of limitation, of cost-consciousness, of regret for the dramatic but rarely profitable Dutch exploits of past decades are striking. The Council always had kept its eye on the bottom line of the ledger, but, down to the 1660s, the Company had been an effective arm of the Dutch war against the Iberian monarchies; it also had sought to increase its profits by attacks on its competitors and to obtain trading advantages in Asian ports by bullying or allying with Asian rulers. But, by the 1680s, the Dutch were at peace with Spain and Portugal, and most of the commercial advantages sought by military and political involvement in Asia had proved elusive. In Ceylon and in several parts of the Indonesian archipelago, political involvement had led to open-ended commitments to territoral suzerainty from which the Company could not or would not withdraw, but elsewhere in Asia every venture had to be justified by the prospects for short-run profit. This was very much true of relations with China. Looking back at the Dutch involvement in that area, the Batavia Councillors saw a depressing continuity.

Dutch Taiwan had not been profitable in its last years. There had been almost no payoff for the Bort expeditions of 1662-1665, either in useful trade privileges or in the monetary compensation the Dutch sometimes thought they had been promised. The expensive Van Hoorn embassy had accomplished nothing.

In planning for the Paats embassy, the Council considered two possible innovations in relations with China and rejected them because of the expense and the small prospects for commercial advantage. One was the establishment of a resident agent in Peking, a possibility that also had been discussed in the instructions for the Van Hoorn embassy. They noted, in addition to the expense and the uncertain benefits, that the person selected would have to be learned in mathematics and other sciences in order to have entrée at the court (as the Jesuits had), and that such people were hard to find at Batavia. Moreover, he and his party would have to be people of exceptionally careful and upright conduct if they were to avoid antagonizing their hosts and thus doing more harm than good for the Company. Paats was not even authorized to agree to a resident envoy if the Ch'ing officials suggested it (a remote chance indeed!), but simply was to say he would report the proposal to his superiors. The Council took almost the same position on the possibility that something might come of Shih Lang's talk about giving Taiwan back to the Dutch. Paats was not to bring the subject up; if the Ch'ing Government brought it up, and if it seemed that the Emperor and people close to him really were intersted in it, he was to avoid cutting off discussion entirely out of fear that they might then offer it to another European nation. He was to say that he had no instructions to discuss the subject, but that he would conjecture that his superiors would not be much interested, because Taiwan had not been very profitable to the Company, and especially because many valuable Dutch ships had been wrecked in its dangerous harbors; but that they might be interested if the Ch'ing would agree to supply as many unarmed Chinese as the Company needed to live on Taiwan and engage in farming and fishing. The council continued to worry about this will-of-the-wisp almost until the departure of the embassy, but the problem was as chimerical as the resident-minister one; by the time the embassy arrived, Taiwan was being firmly incorporated in the regular prefectural structure of the empire.[6]

Vincent Paats, like Pieter van Hoorn, was a recent arrival from the home country with excellent connections, considerable learning, and no experience in Asia. His father, Adriaan Paats, was a distinguished and intellectual Rotterdam regent, patron of Pierre Bayle, Ambassador to Madrid in 1672–1675, one of the Commissioners for the Dutch Company in the negotiations over English claims resulting from the Dutch seizure of power in Bantam in 1682.[7] Only twenty-seven years old, Vincent Paats had arrived in Batavia from the Netherlands just a few months before and immediately had been made a senior merchant and member of the Council of Justice upon recommendation of the Gentlemen Seventeen, and now had volunteered for the ambassadorial post.[8] He had considerably more formal education and abstract interest in government than most servants of the Company. He seems to have studied thoroughly the records of earlier embassies and planned carefully for negotiations in Peking. He conversed in Latin with the Peking Jesuits and with one of their Manchu associates. Later, as we will see, he made some very interesting general comments on the difficulties of negotiating with the Ch'ing authorities. Louis de Keyser was to be second in command and was to take over the embassy if Paats died. He had been the physician of the Van Hoorn embassy, had been in Foochow in 1664, and had been in charge there in 1677–1678, when he had been accused of fraud and corruption.[9] The Dutch seem not to have realized until much too late that the presence in Peking of this physician might have given them opportunities for personal connections within the court or even with the Emperor himself. They might even have been able to leave him behind in some kind of anomalous status comparable to that of the first Jesuit astronomers, but he probably would not have met the Council's very sensible requirement that a resident agent in Peking be an extremely upright and cautious individual. Alexander van 's Gravenbroeck was sent along, promoted from the rank of assistant he had held twenty-four years before to that of junior merchant, because of his knowledge of the country, its language, and its customs. But he probably spoke only

Fukien languages, and knew nothing of China except Taiwan, and nothing of the problems of dealing with a fully developed bureaucracy. Of course, his report of his conversation with Shih Lang was one of the key background documents of the embassy, but he seems to have remained a frustrated marginal figure in it, ignored and treated with condescension by the young patrician and the rather shady old China hand.

At first, the Council had planned to send the embassy via Canton. But on June 19 they changed the destination to Fukien. Foochow had been their most recent fixed trading post (1676-1681); the Ambassador would be able to deny knowledge of a trading voyage that would be going to the islands near Macao; and there would be no Portuguese priests in the area to pose obstacles for them (as there would be at Macao). Although they were interested in maintaining friendly relations with Shih Lang at Amoy, they do not seem to have considered making that city the point of debarkation of their embassy, but would send it to Foochow. Nor did they give much thought to the imperial decision in 1667 that future Dutch embassies must come via Kwangtung.

The embassy left Batavia on July 20 on one ship, not a very good one but the only one available that could get across the shallows into the Min River. It carried a trading cargo primarily of pepper and silver worth a total of only f85,000, presents for the Emperor worth f30,000, presents for various officials from Foochow to Peking worth f31,000, and f19,000 in silver for expenses. Because it was impossible to specify who should receive all the presents and because the Ambassador would have to use his discretion in disposing of presents that had been refused or in buying and giving more and in meeting expenses en route, the scope for account-juggling was wider than in many Company ventures. Paats also asked for and received additional money for extraordinary expenses and a silver table service to use while on the embassy; clearly, he was going to live in the style to which he had been accustomed—if not necessarily in the style in which he probably had seen his father live as Ambassador in Madrid—and, if he could not make the funds stretch, Louis de Keyser could give him some lessons in creative bookkeeping.

The embassy carried letters to the Emperor, the "Imperial Councillors," the Fukien officials, and Shih Lang. The letter to the Emperor congratulated him on the conquest of Taiwan, thanked him for the release of the Dutch captives there, and claimed that the Dutch had been granted "free trade" (which is what they tended to write in referring to any kind of trading privilege), both by the present Emperor and by his father, the Shun-chih Emperor, but "the wars in Your Majesty's lands and also in ours no doubt have prevented the Dutch from enjoying this great favor as we would have wished." It went on to list specific requests for trading privileges very similar to those presented by Van Hoorn in 1667 and by Martinus Caesar, the Dutch headman in Foochow, in 1678: permanent permission to trade in Canton, Chang-chou, Foochow, Ningpo, and Hangchow, and to unload, trade, and depart without awaiting permission from Peking; permission to trade in all goods and with all merchants; possession of a permanent trading lodge in each port where they came to trade; confirmation of the privileges granted in "sealed letters" from the Emperor. Copies of Chinese translations of the main letter and the list of presents for the Emperor were sent along to be turned over to the officials when they asked for them. The letter to the "Imperial Councillors," containing requests very similar to those in the letter to the Emperor, was not to be mentioned in Foochow, and if, on discreet inquiry in Peking, it seemed likely that an attempt to deliver it would cause difficulties (as such an attempt had for Van Hoorn) and not open any new channels of communication, it was to be brought back undelivered. The Council's instructions for Paats authorized him to make additions to or deletions from these requests if changed circumstances made it advisable, to accept limitation of trade to one or two ports, and to agree to send embassies at regular intervals if the Ch'ing officials insisted on it. Their instructions on the possibility of a resident minister in Peking and on that of an offer to return Taiwan to the Dutch have been discussed above. The Ambassador was to confirm to all Chinese customs and ceremonies "as far as equity allows," and use great discretion and tact. The presents for the Emperor included coral, amber, mirrors, European and Indian fabrics, a table clock, a copper lantern, a copper candelabrum, glassware, cloves, sandalwood, bottles of oil of cinnamon, oil of clove and other oils, elephant tusks, fine swords and guns, three telescopes, three little models of ships, and two bottles of Spanish wine.[10]

The embassy ship arrived at Ting-hai in the Min River estuary on August 25. The weeks that followed were full of delay and frustration for the Dutch, caused by the treacherous channel, the bureaucratic reporting requirements for a tribute embassy, and the antagonism between the provincial officials and the new Maritime Trade Commissioner or Hoppo. At first they were visited every day at Ting-hai by officials seeking information on the measurements of the ship, the number of people and quantities of trading goods and arms aboard, and a full list of the presents to the Emperor. They gave most of the information, but said they would not be able to give a full list of the presents until they all were unpacked. They had a list, of course, but would want to check to see that nothing was broken or missing. These officials also asked repeatedly why they had not sent an embassy after the prescribed eight years, and why they had come via Fukien when they had been ordered to come via Kwangtung.[11]

On September 4, having received permission from the Hoppo to bring their ship into the river, they got it stuck fast on a sandbank, and the Ambassador and his party went on up to Foochow on junks. Representatives of Governor-General Wang Kuo-an and Governor Chin Hung soon came to ask on whose authority the Dutch ship had tried to enter the river. They made it clear that there was no love lost between their masters and the Hoppo, and insisted that the embassy was completely outside the Hoppo's jurisdiction. Worse, the Ambassador could not be received by the high officials until the presents to the Emperor had been brought up, and apparently he could not be given regular lodgings until he had been received. So Paats was left fuming for four days in his cramped, hot junk. Some Chinese musicians sent by the officials to entertain him one afternoon did not help much. Finally, after the presents had been brought up, he was received very cordially on September 10. The Dutch were astonished that there was no mention of the long delay and mis-routing of their embassy; apparently the officials had full enough responses from the earlier questionings by subordinates to be able to make full reports to Peking. They were offered, and accepted, Keng Ching-chung's old mansion for their use, rent-free, during the embassy's stay. Chin Hung would meet the Dutch only in the "full assembly" of all the officials, including Wang Kuo-an, and, when these two officials received their letters from Batavia, they immediately handed them to each other for inspection, and the letters then were passed around among all the officials present. The high officials later told the Dutch that new imperial orders forbade officials to receive an ambassador alone or in their private residence before the ambassador had been received in Peking. The general principle of no private communication between provincial officials and foreigners was clear, but I have found no reiteration or specification of it in the 1680s.

On September 13 and 14, Wang and Chin sent placards making it clear that trade now was open to all foreigners, whether or not they came to present tribute, but also that the Dutch embassy posed new problems. It had come after nineteen years instead of eight, and via an unauthorized route. Under the new regulations, 3 ships bringing an embassy were to be exempted from tolls. But they would not be allowed to trade until approval of their embassy was received from Peking; probably this was because the application of this rule to them would depend on imperial approval of their tardy and mis-routed embassy. (It had been decided in 1684 that future Siamese embassies would be allowed to trade in Canton without awaiting approval from Peking, and that they would be allowed to send home the ships that brought the embassy party and have more ships come to take him home, but the decision to apply these precedents to the Dutch would have to be made in Peking. A decision earlier in 1685 had exempted embassies from the new tolls, but this could not be applied to the Dutch until their embassy was approved.[12]

The Dutch finished unpacking and checking the imperial presents and turned over a list of them on September 18. On September 20, Wang, Chin, and the Hoppo came to see the presents:

> Those of the suite who understood music and the playing of instruments showed their arts at the request of Their Highnesses, and made a fine harmony with their agreeable voices, which caressed their ears and pleased them exceptionally, as they declared to the Ambassador. One of the servants, quite accomplished in dancing and all kinds of grimaces and tricks, increased their amusement and pleasure not a little, and they showed by their laughing faces their astonishment at his nimbleness.[13]

On September 28, the Dutch turned over a copy of the letter to the Emperor and a list of the people who would go to Peking. Previously a subordinate official had explained that the Governor-General had to see the letter to make sure that there were no mistakes in the titles used in it and to see if anything bad had been written about anyone. The Dutch were assured that the submission of these documents cleared the way for permission to trade and for departure of their ship, and that placards would be sent permitting trade as soon as they turned in a full list of the trading goods they had brought. The Hoppo had been demanding such a list since September 15, along with information about the measurements of the ship, the number of men aboard, and its munitions. The Dutch did not want to disclose the full quantities of their goods until they were permitted to trade, fearing quantitative restrictions on trade like those recently adopted in Japan. Twice, on September 22 and 26, the Ch'ing authorities agreed to accept a list of smaller quantities than the actual cargo, but, from the 28th on, they again demanded full disclosure. Apparently no full list had been sent and no placards permitting trade had been received by October 15, when our only surviving day-register ends, and it seems that trade was not opened until November 11, probably after approval had been received from Peking. The authorities may have gotten much of the information they needed about ship, crew, and weapons by sending a low-ranking official out to inspect it at its anchorage.

The Dutch day-to-day record down to October 15 is full of

Ch'ing assurances that placards would be sent; arguments about phrases in Dutch requests for them; difficulties in obtaining passes for boats to move up and down the river; and efforts to offer presents to officials who were eager to receive them but extremely wary of being caught doing so. The arrival in the Min estuary of a Dutch ship that had lost its voyage to Nagasaki led to more demands for information and to orders that it depart. The similarity to Van Hoorn's frustrating autumn in Foochow in 1666 is striking. If the Ch'ing Empire in general was somewhat more open to foreign contact than it had been nineteen years before, in Foochow this was offset by the new bureaucratic reporting requirements for maritime trade and taxation of it and by the new and edgy relation between the Hoppo and the provincial officials; no one could be as free and easy with the foreigners as Keng Chi-mao and Chang Ch'ao-lin had been in 1666.

According to the Dutch, the approval of the embassy that was received on November 11 also authorized the departure of their ship without awaiting the Ambassador's return from Peking, despite contrary precedents (especially those of the Van Hoorn embassy), because the ship was damaged (and thus had to return to Batavia for repairs). The Siamese precedent mentioned above probably facilitated this decision. Thereafter, trade apparently moved rather well: there was no indication that it was confined to a few favored merchants, and it was conducted on cash terms under close official supervision, quite similar to the trade of the Van Hoorn embassy at Peking. The high officials did not participate in the trade directly, and any connections they had with particular groups of merchants were kept concealed from the Dutch; reportedly, they feared that the Hoppo would report their involvement in trade to Peking.[14] Thus, a nice synergy of tribute-system precedents and bureaucratic rivalry in Foochow had produced uniquely advantageous conditions for Dutch trade in Foochow, and the Dutch had nothing to sell except a small cargo, mostly pepper.

On November 15, Alexander van 's Gravenbroeck was sent to Amoy to deliver the letter and presents from the Batavia authorities to Shih Lang, and to see if Shih could help them obtain favorable consideration by the court of the Dutch requests.

Shih and the other officials on Amoy greeted him cordially, and said the Dutch should come to Amoy to trade. Shih said he would write to the court on behalf of the Dutch. Van 's Gravenbroeck returned to Foochow on December 10. That was all that came of Van 's Gravenbroeck's suggestions and Shih's boasts of influence two years before.[15]

For the rest of the embassy Dutch records are fragmentary, and some points can be found only in the records of a quarrel between Paats and De Keyser on one side and Van 's Gravenbroeck on the other, which continued after their return to Batavia and fills over 200 pages in the Company's records.[16] Many people found Van 's Gravenbroeck hard to get along with, and he had not been of much use as an interpreter, but apparently the real root of the trouble was his unwillingness to condone the corrupt practices of his superiors. For example, they recorded as expenses the full value of all presents offered, even if they were not accepted; in that case, Paats and De Keyser probably sold the goods for their own profit. Van 's Gravenbroeck refused to sign resolutions covering up this practice. After the embassy returned to Foochow, Paats and De Keyser placed him under arrest and searched his personal papers, apparently fearing that he had kept records of their practices, but found nothing. After lengthy consideration, the Batavia Council decided on 5 March 1688 that Paats owed the Company f20,096, including the imperial present of 300 taels for the Governor-General, which he had appropriated, but Paats had left for the Netherlands, and I do not know if he finally had to pay or not. Paats may have taken the missing day-registers of the embassy home with him because he intended to publish them or because they contained too much detailed evidence on his financial manipulations, but I have not been able to find any trace of them in the Netherlands.[17]

STRANGE ENCOUNTERS IN PEKING

The embassy party left Foochow on 17 March 1686. I can think of no reason for the long delay except the desire of the Ch'ing officials not to subject southern embassies to the northern winter and to avoid the difficulties of winter travel on the Grand Canal. At Hangchow, late in April, they investigated prices and tried to get information about Chinese overseas trade from that area. The officials there told them that foreigners were permitted to trade only in Fukien and Kwangtung, but, if any Dutch ships came to Chekiang, they would recommend that they be allowed to trade. The Dutch tried to investigate the estuary at Hangchow, but the Chinese boatman whom they had hired fled, apparently fearing punishment for helping foreigners spy out routes. In Shantung, they had the usual difficulties with the Grand Canal, being forced to wait twelve days at Lin-ch'ing because there was not enough water for the locks.[18]

The embassy arrived in Peking on July 31, and was received at an ordinary audience on August 3. According to Van 's Gravenbroeck, a translation of Paats's "oration" was read at the audience by Father Verbiest, and the Emperor spoke to Paats, with Verbiest interpreting. This could have been at the brief interview within the T'ai-ho Hall that frequently followed the formal audience, or at a later audience elsewhere in the palace. It is interesting to recall that Verbiest apparently had read aloud a translation of the "royal letter" Pereira de Faria had brought. But all this is very vague in our Western sources, and never became part of the statutory routine of a tribute embassy. Van 's Gravenbroeck did not know what they said, because Paats and Verbiest spoke "French and Latin." When they had returned to their lodgings, Van 's Gravenbroeck asked Paats what the Emperor had said. "His Excellency winked to his cousin Adriaan van Heemskerck, and answered only that 'the Emperor had said that he had understood that His Excellency Mr. Paats was a man of great learning, especially in the mathematical arts, and that His Majesty had asked His Excellency if His Excellency knew in what latitude Holland lay,' but nothing more, so that no further opening was given."[19] This was, of course, in the years when the Emperor was learning a good deal of astronomy and mathematics from the Jesuits and was taking great pleasure in making astronomical observations and calculating latitude himself when on tour beyond the Great Wall.[20]

Thereafter, the embassy party went through the usual routine.

Paats reported that they had been treated better than other ambassadors, had been granted some kind of special banquet, and the Emperor had spoken to them affably and out of curiosity had asked all sorts of questions; it sounds as if this was at a banquet in the imperial presence, but we cannot be sure. At such a banquet or at some other imperial audience of which we know nothing at all, some of the musically talented members of the Dutch party seem to have had an opportunity to perform before the Emperor, which eventually led to two of them being left behind in Peking. The Peking Jesuits visited the Dutch frequently; reportedly the Emperor had ordered them to keep closely in touch with the Dutch. Sometimes they brought along Chao Ch'ang, a member of the Imperial Bodyguard and a key intercessor with the Emperor for the Jesuits in these years. Chao occasionally came to visit the Dutch alone, and became a favorite intermediary in their efforts to learn what the court would decide about their requests and to secure a favorable decision. He sometimes conversed with Paats in Latin, and was said to be interested in Western science.[21]

About August 14, Paats presented a document repeating the basic requests in the letter from the Batavia authorities to the Emperor, making it clear that "free trade" meant trade "without payment of tolls, as previously," and requesting that the Emperor order his Councillors to discuss orally with him conditions of trade and the interval for embassies. The request for exemption from tolls was his own addition; the Batavia authorities had had very limited information on the new taxation of maritime trade when they wrote the letters and instructions for the embassy. Paats continued to plan for negotiation and amplification of the requests in face-to-face discussions. He thought it might be possible to obtain, if not total exemption from tolls, a reduction of them, or some kind of commutation to a fixed sum from each ship or each year's trade, which would free Dutch trade from the constant petty vexations of dealing with the toll collectors.

There was no explicit Ch'ing response to most of these requests. The requests for permission to trade with all merchants, in all goods not explicitly prohibited, and to unload, trade, and depart without awaiting approval from Peking had been more or less met by the opening of trade and the dispatch of the maritime trade commissioners. Of the five ports in which they wanted to trade, Ningpo and Hangchow were not yet open to foreigners. There was no response to the request for exemption from tolls. The court did respond to, and reject, the request for permanent trading lodges. This definitely was beyond the limits of the new order on the coast. In it, all foreigners might come, but none might stay the year round, except at Macao. The Dutch had stayed over year to year from 1662 to 1666 and from 1676 to 1681, but those stays had been ad hoc arrangements, with occasional prospects of military cooperation and no taxation of trade. Now there was a different pattern, in which foreigners might come and their trade would be taxed, but there was no reason for them to stay over. The basic principles of this order were not violated until the Opium War. Thus, despite the rather relaxed and cordial reception of this embassy in Peking, the context of coastal policy changes made it certain that it would be able to gain very few concessions for the Dutch Company. Paats later suggested that the court had rejected the request for permanent trading lodges in part because the conflict with the Russians had turned it against all Europeans. Certainly the Russian problem loomed very large in Peking at this time, but I doubt that it had anything to do with the decisions on the Dutch. The Russians were seen as part of the Inner Asian world, with little connection to Europeans who came by sea. The coastal world had a new set of rules that had to be applied impartially to all foreign traders.[22]

On August 20, having learned that the request for permanent trading lodges was to be refused, Paats submitted a new written request that the Dutch be allowed to trade in all ports and establish trading lodges, and asked that they be informed as to how often they should send embassies to Peking. Apparently, he never was allowed to discuss his proposals orally with the officials, except for two discussions concerning intervals of tribute embassies on August 23 and September 8. On the latter date, Paats pointed out that now others were enjoying the same privileges as the Dutch without sending embassies at all, and declared that the Dutch would not be bound to send embassies at fixed intervals unless they got some advantage over their competitors. But in the final decisions the Dutch tribute period was reduced to five years, "and it was even considered as a favor." Chao Ch'ang assured the Ambassador that he had spoken to the Emperor on behalf of the Dutch request for permanent lodges, and that it would be approved even if the Board of Ceremonies recommended against it. Various high officials sent emissaries to tell Paats permission for a lodge could be obtained if enough presents were given. One fragmentary accounting says that presents were offered to Board of Ceremonies officials, that presents worth perhaps 500 taels were accepted by lower-ranking Board officials but none by higher.[23] A brother of the Emperor was given a fine gold clock. Chao Ch'ang accepted presents worth about 150 taels. All this makes an interesting contrast to the absolute refusals of presents in Peking in 1667. Paats rejected a proposal that he make a single present of 6,000 taels to be distributed among all the necessary officials, pointing out in his report that the sum was far too small to persuade so many officials to risk life and honor. Chao Ch'ang and others continued to "feed him with hope," but, early in September, he learned that the requests would be refused. He then made a last request, that two of the ships that came to take the Ambassador away be exempted from tolls. This was granted, on the ground that three ships were permitted for an embassy and he had come with only one. On September 14, the Dutch received the edict to the "Holland King," the imperial presents, and communications from the Board of Ceremonies transmitting the imperial decisions: the Dutch were to send tribute every five years, via Fukien; they might come to trade in either Fukien or Kwangtung but must all depart at the end of the trading season; they should send smaller quantities of tribute goods, and only certain specified kinds; two of the ships that came to take the Ambassador away were to be exempted from tolls. Paats asked to receive these documents unsealed, and this was permitted; a sharp contrast to the Van Hoorn embassy, which had been forbidden to break the seals within the empire. The Dutch also took along an edict from the Emperor to the Russian Tsar; the Emperor had received no reply to previous edicts calling on the Tsar to keep his subjects from disturbing the borders of the Ch'ing Empire, and was anxious to find a new route by which such edicts could be sent. The Ch'ing authorities asked the Dutch to carry any reply the Tsar might send, or to send word if he refused to reply.[24]

Later, the Batavia Council accused Paats of having trusted the Jesuits too much. Paats had expressed his gratitude in a letter to the Fathers and in his reports, but he also insisted that he had asked to use his own interpreter at his audience and this had been granted, but that he had been forced to accept Verbiest at the last minute.[25] I do not see any evidence that the Jesuits really hindered the Dutch in any way. Their associate Chao Ch'ang seems to have been very helpful, but his optimism about the concessions the Dutch might obtain from the court may have been partly motivated by hope of more presents for himself and others. The Jesuits certainly did not want the Dutch to become so firmly established in the China trade that their competition would do further damage to Macao. But they could not go too far or be too obvious in their maneuvering, since a small but significant part of their influence at court was the result of their reputation as reliable and reasonably impartial translators in dealing with the Dutch, the Portuguese, and the Russians.[26] They did try to limit the credit the Dutch might gain for delivering the edict to the Tsar. Verbiest recommended to the court that Philip Grimaldi, S.J., carry to Europe another copy of the edict, hoping that the Jesuits would turn out to be the better channel of communication. Actually, it seems that the copy carried by the Dutch reached Moscow a few months before the one carried by Grimaldi.[27]

The embassy party left Peking on September 17. It left behind in Peking, at the urging of the Jesuits and apparently because the Emperor had heard and liked their music, one Frans Flettinger, a German sergeant of rather dubious reputation, onetime steward of the embassy, who played the violin, and a Javanese slave who played the harp. We have only a short summary of the diary Flettinger kept in Peking, but even that is astonishing, with its hints of court anxieties about the Russians and its vignettes of the Jesuits oscillating between hope and despair in dealing with the frivolity and contempt of much of the court and with the Emperor's shifts from delight with their science and gadgetry to sharp questioning of their most basic beliefs.[28]

The embassy party reached Foochow on November 12, having been moved along day and night most of the time. Someone told Paats that Chao Ch'ang's influence was responsible for their being moved along so quickly, and the Dutch sent him a fine clock he had asked for.[29] Four Dutch ships had arrived off Foochow on August 23 and 24. The Dutch had already been given permission to trade and had been granted exemption from measurement dues and tolls for two of their ships, but a price agreement was not reached until some time in January 1687. Trade was wound up by March 1, and all the ships sailed for Batavia on March 10. The Ambassador reportedly had been treated very coolly by the officials after his return from Peking, and he left without a farewell audience. The embassy was welcomed back to Batavia on May 20, with cannon salutes, musket volleys, and a little procession bearing the Emperor's letter and presents to the residence of the Governor-General.[30]

It is interesting that the Paats embassy had passed up or made minimal use of some of the best chances the Dutch ever had to establish personal contact with a Chinese Emperor. First, at this time there was a small craze for Western music in the K'ang-hsi court, and Father Thomas Pereira, S.J., was busily building musical instruments, demonstrating his own talents, and teaching the elements of Western music to members of the court and even to the Emperor himself.[31] Dutch music had made quite an impression in Foochow and in Peking, but Flettinger and the slave were left behind as an afterthought, at Jesuit instigation, and neither gave much promise of being a trustworthy agent of the Company at court. Second, the Jesuits were taking great advantage of the court's penchant for clocks, fancy fountains, and mechanical toys. The Dutch included a clock, some telescopes, and ship models among their presents to the Emperor, and gave two more clocks to the Emperor's brother and Chao Ch'ang, but did nothing to follow up on these opportunities. Third, a gift of up-to-date firearms might have been welcome in view of court anxieties about the Russians. Fourth, at this time the Emperor had told the Jesuits he would like to have a Western physician at court, and the Jesuits, realizing that a physician would have far more personal entrée to the Emperor than an astronomer, were trying to find a suitable person somewhere in the Society's Asian establishments, but were not having much luck.[32] Dutch physicians had helped cement relations with the Fukien officials in the 1660s, and Louis de Keyser, the second in command of this embassy, was a physician, but the Dutch do not seem to have been aware of this opportunity while they were in Peking. Later, Flettinger wrote that the Emperor had learned that De Keyser had treated various officials and seemed inclined to make him his personal physician. In the relatively open atmosphere of Peking in the 1680s, it is *conceivable* that the Dutch could have established a small resident group of musicians, clock-makers, gunsmiths, and so on, centered on a physician attached to the court. But, as we have seen at the beginning of this chapter, the Batavia authorities were not at all interested in having a resident agent in Peking. Paats also had learned from the Jesuits how much difficulty they were having in attempting to gain some small favors for Macao, which strengthened his conviction that a resident agent would not be of much use.[33]

The failure of the Paats embassy ended the involvement of the major European maritime trading powers with the tribute system for a century. As Paats pointed out, the English trading at Amoy might have sent an embassy if the Dutch had profited by theirs. The French, with a strong position in Siam, were said to be showing some interest in the China trade. Flettinger reported that a French Jesuit had come from Siam to Macao to investigate trade and to learn if the Dutch embassy had been successful.

But the irrelevance of tribute embassies to maritime trade was apparent even to some of the Ch'ing authorities. The maritime trade commissioner at Amoy told Van 's Gravenbroeck that "from now on it would not be necessary to send any embassy, for the Emperor had absolutely opened trade to all foreign nations, and those that wanted to come with an embassy would be exempted from tolls for three ships."[34] And, according to Flettinger, who got his information from the Peking Jesuits, when the Emperor learned that the Paats embassy had cost the empire 27,300 taels, he said that "these burdens fell too heavily on the land, and he wished they [embassies] would all stay away."[35]

Nevertheless, Paats had learned that, as an obstacle to negotiation in Peking and a shaper of Chinese views of foreign relations, the tribute embassy was still a very powerful institution:

> . . . it can be concluded that the rights and dignities of ambassadors are not understood by the Tartars as they are by all Christian princes (although they [the Tartars] stand out far among the heathens for their customs), and that they do not want to discuss matters with any foreign ministers, but decide everything according to their own opinion, and establish their decisions as law for the foreigners; which the word *chincon* [*chin-kung,* tribute] implies in its original meaning, and that those who abase themselves to greet the Emperor with presents, must prepare to receive his commands, and to obey them when they frequent the empire; which is especially confirmed for the ambassadors of all other nations, such as Siam, Eynam [*sic*: Annam?], Cochin China, Tonkin, Koreans, and others, who must make amends whenever they do not respect the fixed time for the chincon or their presents are lacking in quantity or type. And all must use the Emperor's seal, which, however, being a sign of the greatest contempt and abasement, is not required of the Europeans, who are considered by the Tartars, although far beneath themselves, far above all other peoples, and are recognized as independent.[36]

But Paats had found organization as important as ideology as an impediment to negotiation at Peking:

> Few great officers except these favorites have access to the Emperor, and even these, in order not to be suspected of having taken presents, dare not speak on behalf of foreigners except with the greatest circumspection and when the Emperor gives an opening for it; also, for the same reasons, the provincial officials dare neither to speak nor to write on behalf

> of foreign embassies. Therefore these presents were not projected and given on behalf of the Honorable Company for any other reason than the offerings that are made by various Indian peoples to their evil spirits, not for the sake of some good, which is not in their nature, but in order not to receive any evil . . .
>
> That the present Emperor, standing out far above his father in his regard for foreigners (as is affirmed by the reception and honor accorded us by his government, superior to all other ambassadors) refused the Honorable Company a dwelling even under payment of the tolls, is only to be attributed to the fearful nature of that Prince, who, in order not to place all the hate of bad results on himself, seldom decides on the basis of his sovereign power against the counsels of the respective colleges, from which, being a many-headed, divided, and interested gathering, no decision to the benefit of foreigners . . . is to be expected, unless it has been purchased in advance for a considerable sum of money, according to the importance of the matter.[37]

Considering the brevity of the author's acquaintance with China, the second passage shows admirable insight into some of the effects of late imperial status-quo despotism on foreign relations. The former brings us right back to tribute ideology and institutions, and so do the Chinese documents. The *Veritable Records* record only the arrival of the embassy in Peking, its imperial audience, the edict of praise to the "Holland King," and the decision to have the Dutch take an edict to the Tsar.[38] One unofficial source makes much of this last item as an example of the theory of "using foreigners to control foreigners."[39] The edict of praise betrays no hint that the recipient was anything other than a more distant analogue to the King of Ryukyu: "You, King of Holland, live in a distant region . . . That you have sent an envoy with a memorial and tribute shows a depth of loyalty that can be commended." In other bureaucratic sources and even in some private ones the picture of the tribute system as one of unilateral and minute regulation is maintained in records of the changes in the rules on the frequency and route of Dutch embassies (including the remarkable assertion that the Dutch, having been permitted to pay tribute every eight years, "moved by the Imperial Benevolence, had again requested the fixing of a [shorter] period"), full lists of the presents the Dutch received, those they brought, and the more modest list of goods they were told to bring in the future.[40] The Dutch had translated these orders, but there is no hint that they understood that they were gestures of imperial benevolence within the stereotypes of a ceremony-centered system, or that their presents had been perceived and accepted within a symbolic framework that claimed continuity at least from those hounds of Lü in early Chou. Two sources quote an extraordinary "translation" of part of the "tribute memorial of the Holland King" that has nothing to do with the Batavia authorities' letter to the Emperor, and demonstrates better than anything else the continuing victory of appearance over reality in the tribute system as the European trading powers disappeared from it:[41]

"Every ball of mud and foot of earth for foreign countries is but a particle of dust flown from the Middle Kingdom. Every spoon of water, every hoofprint filled with water in strange lands has its origin in the falling dew of the Celestial Household."

chapter six

The Survival of Ch'ing Illusions

This book began with a description of Pieter van Hoorn participating in the core ceremony of the tribute system, a description that depends largely on the Dutch sources but fits well with what Chinese regulations tell us about tribute embassies and audience ceremonies. Beyond this and a few other basic ceremonies, the pictures given by Chinese and European sources differ strikingly in emphasis. The bureaucratic system and millennial continuity so clear in the Chinese sources never were sensed by the European participants. The European sources offer no alternative unifying vision; to sit in the archives of Rome or Lisbon or The Hague or Goa and look for the tribute system is to court frustration and puzzlement, sometimes to feel lost in anecdote and marginal context, but eventually to end with a clearer picture of the system and of European participation in it. Only the European sources enable us to see the importance of non-statutory events in Peking and of the handling of the embassy away from Peking in the total picture of an embassy. Only through detailed study of European backgrounds and contexts can we see clearly the accidental and multiply caused nature of European involvement in the tribute system.

Nor is the value of these European sources limited to new insight about the tribute system. We have had some remarkable glimpses of the curious, willful young Emperor and his court, and come to understand a little more of the way in which he used his treatment of the Jesuits and the Saldanha embassy to emphasize his break from the Oboi Regency. The great extortions from Macao are the best-documented case I know of the corruption and tyranny of the coastal evacuations. We also find a good deal of evidence in support of some generalizations about early Sino-Western relations that are not applicable to tribute embassies alone. Perhaps the most important of these is the idea that, when one wants to understand a shift in Chinese policy toward a foreign power, one must always pay attention to shifts in the politics of the court and of a relevant provincial capital, as well as to the usual ideological factors and the political and economic goals to which the policy seems to have been oriented. The need to keep domestic political factors in focus is fairly widely understood among historians of foreign policy, but it is striking to see how often in these studies the timing of a decision can be explained only in this way: the revocation of the Dutch biennial privilege in 1666; at least four turning points in Macao's perils in the 1660s; the final acceptance of the Saldanha embassy; possibly the final fruition of the concessions obtained after the Pereira de Faria embassy. I have also suggested the relevance of domestic political shifts at several points in my brief summary of Ming foreign policy in Chapter 1.

The hypothesis that internal political factors were more likely to dominate foreign policy-making in China than in many other polities can be fitted quite readily into the basic picture of the Chinese diplomatic tradition I began to develop in *Pepper, Guns, and Parleys;* in a large, single-centered polity, the great bulk of political issues, political experience, power imperatives were internal, in a way they might not be for, say, Venice or the Netherlands. I also pointed out previously that it was partly because of the character of domestic politics that there was so little specialization of personnel or bureaucratic bodies in foreign relations. This also is clear in this study; embassies were in the hands of the high provincial officials, the Board of Ceremonies, and an occasional inner-court figure like Chao Ch'ang.[1]

In my earlier book, I also sketched a typology of sources of irritation and conflict in early Sino-Western relations. Many of the same irritations can be seen here: European annoyance at the restrictiveness of Chinese bureaucratic routine (Van Hoorn, Paats); Chinese disinclination to explain their own procedures to the Europeans (especially striking in the Van Hoorn embassy ship departure problem); European disregard for the importance of courtesy and form in dealing with the Chinese (Saldanha's refusal to seek audience with the Canton high officials). As I suggested, if we want to keep in focus these aspects of Sino-Western conflict, we must not focus too closely on conflicts of basic values and views of foreign relations, but must see foreign relations as contacts on various levels between many features of two cultures, two political and social systems. But there were many sources of irritation which, while not limited to the tribute embassy context, when they appeared *in* that context can be seen as reflecting the clash between the basic values and world-views of the tribute system and those of Western formal diplomacy. In the embassies studied in this book, there was almost no explicit conflict over these fundamentals. Rather, basic differences can be seen reflected in a variety of clashes of unexamined expectations. In 1666, the Foochow authorities seem to have expected that the Dutch ships would stay over; in 1670, Saldanha expected to be allowed to take his whole suite to Peking. The Europeans did not expect to trade in the capital in connection with an embassy. The Ch'ing authorities expected embassies to be largely ceremonial occasions, while the Europeans anticipated occasions for negotiation and binding, documented agreement. The minute bureaucratic control of an embassy, the strict control of its movements and its contacts with the Chinese, the delays caused by bureaucratic procedures were perfectly normal to the Chinese, but unexpected, irrational, and extremely frustrating to the Europeans.

These sources of conflict, however, here linked to the embassy institution and thereby to basic clashes of values and world-view, also can be seen in many circumstances where there was no such link. Bureaucratic restrictiveness and delays, for example, probably were the most basic sources of European irritation in relations with China, occurring in many situations that had nothing to do with tribute embassies. In my first book, I argued that we could not keep in focus all aspects of the Chinese diplomatic tradition, all sources of conflict, if we began by calling all of the Chinese diplomatic tradition the "tribute system." But this leaves open the question of the relation of the more complex phenomenon I call the Chinese diplomatic tradition to that part of it which was a complex of regulations and institutions directly associated with tribute embassies, for which the term "tribute system" is perfectly apt.

It should be remembered that the tribute system as a system cannot be traced back farther than the Ming. In Ming times, it embraced all aspects of relations with all foreign countries, in theory and to a large degree in practice. In Ch'ing times, it maintained this comprehensive character for relations with Korea, Ryukyu, Vietnam, and Siam, but not for those with Inner Asian peoples (including Russia) or maritime Europeans. And, when we see our four embassies in their contexts of ongoing relations in Canton and Foochow and of Ch'ing internal politics, it would be hard to argue that the tribute system was among the most important determinants of the course of Sino-Dutch or Sino-Portuguese relations in the early K'ang-hsi reign.

Vincent Paats had remarked that the presents his embassy had given to various officials were like those some Indian peoples give to idols, not to receive favor but in order not to receive evil. The metaphor is peculiarly apt for the two Dutch embassies. If Van Hoorn had not come in 1666, it is very likely that the Dutch would not have been allowed to trade at all that year. I am less sure about what would have happened in 1685 if the Paats embassy had not come; some trade might have taken place in the islands near Macao, but I think it unlikely that any Dutch trade in a Chinese city would have been allowed. Other effects of the tribute system on Sino-Dutch relations are less important or less clear. The costliness of the embassies and their manifest uselessness as channels of negotiation certainly contributed to the consolidation of negative attitudes in Batavia toward continued relations with China. The review of precedents on the Dutch and the resulting cancellation of their biennial trading privilege might not have taken place if the Van Hoorn embassy had not come, but that is purely speculative. Although the Dutch failure to send an embassy in 1664 or 1665 had irritated the authorities in Foochow, and some of the decisions on Sino-Dutch trade and cooperation from 1662 to 1665 were couched in the terminology of the tribute system, the dynamic of the rise and fall of Dutch-Ch'ing entente in these years was centered not in the tribute system but in changing Ch'ing perceptions of the Dutch as naval allies, valuable in time of need but too dangerous and unmanageable to be allowed to stay around once the need had waned. In 1677-1680, Dutch trade was permitted and Dutch naval assistance was sought, despite the fact that a Dutch embassy was long overdue, and, as far as I can tell, without reference to the concepts and institutions of the tribute system.

The Portuguese case is a bit more complicated. The arrival of the Saldanha embassy helped assure Macao's survival in the summer of 1667, and the many signs of imperial favor to it in Peking in the summer of 1670 enhanced the prestige of the Portuguese and of the Peking Jesuits. But if we want to play another counter-factual game and ask what would have happened if no embassy had come, I think it very likely that Macao would have survived because of the power shift at Canton in 1667, and that K'ang-hsi's favor to the Jesuits and their allies in the bureaucracy would have been sufficient to shield Macao from later attacks. And, as we saw at the end of Chapter 3, the Saldanha embassy obtained no concrete or documented concessions, and, as soon as the Canton officials were sure it was not going to produce any further reprimand or investigation, they resumed their extortions.

The embassy of Bento Pereira de Faria *did* obtain concrete, documented concessions, although full implementation of them was postponed until the Cheng regime was driven from the coast and even then may have been facilitated by the fall of the Shang regime in Canton. The success of the embassy despite its improvised nature and bogus documentation cannot be entirely explained by the splendid present it brought; lions were not unknown in Peking, and an embassy bringing only one present was decidedly irregular. The acceptance of this embassy probably owed something to a waning inclination to apply all the bureaucratic categories of the tribute system to all foreign relations, already visible in the failure to establish any mandatory frequency of Portuguese embassies after the Saldanha embassy. Its success certainly owed a great deal to Verbiest's influence and persuasive skills; it is even possible that he could have obtained the same concessions for Macao even if no embassy had come.

Even if embassies were not structurally or causally central to Sino-European relations in this period, they still might have fulfilled substantial functions. One function they did not fill, as we have seen, was that of channels of communication and negotiation. The obstacles to negotiation in Peking were largely rooted in the tightly regulated, ceremony-focused routine of an embassy in the capital, but also in some very practical political causes. Policies to be implemented in Foochow or Canton would be based on proposals made by the high officials in those places, and only exceptionally on initiatives from Peking. As we can see from the favorable reception and relative success of the two Portuguese embassies, in these circumstances nothing the sending sovereign could do made as much difference in the success or failure of an embassy as the presence or absence of resident friends in Peking to guide it, get in touch with officials on its behalf, and carry on its negotiations after it left Peking. It would be interesting to see if such connections also were important between embassies and eunuchs from the same frontier people in the Ming, or between Inner Asian embassies and Mongol officials or Mongol and Tibetan lamas resident in Peking in Ch'ing times.

There were a number of other functions that were attributed to the embassy in Chinese lore or were plausible uses of its elaborate codification and bureaucratic management, but they do not seem to have been applied in any coherent way in the handling of these four embassies. The use of the embassy institution to give foreigners incentives—titles, gifts, bureaucratically controlled trade—to do as the Chinese court wished them to do seems to have had almost no place in the management of these embassies. Only one of the four traded in Peking. The Ch'ing court does not seem to have thought it was "enfeoffing" the King of Portugal or the "Holland King." There is no indication from either Chinese or Western sources that Ch'ing officials tried to use any of these incentives to make the Europeans more tractable. (It is hard to imagine how Macao, scarcely surviving, could have been any more tractable.) Of course, unilateral bureaucratic administration of trade and its manipulation to bring foreigners to heel can be seen in Ch'ing policy toward the Dutch on the coast in these years, and toward others at Canton, Kiakhta, and elsewhere in the eighteenth century. No doubt the background of the administration and manipulation of tributary trade helped shape these policies, but in themselves they had nothing to do with tribute embassies. I do not know of a case of political manipulation of incentives offered within the tribute system during the Ch'ing dynasty; I look forward to someone finding one, documenting it, and thus adding some new twists to our understanding of Ch'ing diplomacy.

Second, the sharply defined levels of banquets, rations, and gifts for tribute embassies would seem to have been admirably suited to the indication of imperial favor for one sovereign over another by granting the favored one a higher level of rations, more presents, a banquet in the imperial presence, or what have you. Non-statutory aspects of the embassy offered even more interesting opportunities for the communication of favor: an imperial appearance to view the presents, a summons to an informal audience, permission to visit the imperial pleasure parks. Saldanha's Jesuit mentors made much of every sign of such favor. Although they no doubt were a bit carried away by their own enthusiasm and their desire to emphasize their success in aiding the embassy, I think they were reflecting something that really was there in the way Ch'ing courtiers looked at an embassy; courtiers had to be alert for signs of imperial favor to individual ministers, and there is no reason to think that they would not be watching for such signs toward foreign countries as well. We have one striking case of imperial favor in application of the formal categories, in the larger imperial presents bestowed in connection with the Pereira de Faria embassy. Non-statutory, informal signs of favor to the Portuguese embassies can be seen, but they can be matched for the Dutch embassies: the Emperor's coming to view the Van Hoorn presents; Paats's musicians performing for the Emperor. These examples suggest that the Emperor's private tastes and curiosities might be as important as any politically motivated communication of favor in determining the informal, non-statutory favor shown to an embassy.

Finally, there is one other function of the embassy which had been central to Chinese ideas of foreign relations since Han times, in which the Ch'ing authorities were very successful in our four cases and in almost all their foreign relations down to the nineteenth century. This was the preservation of the forms, the appearances, of imperial suzerainty over all foreign rulers. The traditional Chinese view of foreign relations did not require the assertion or actualization of such suzerainty outside China, but did insist that all foreign envoys coming to China would have to formally acknowledge it in the correct ceremonial and documentary forms. The presents sent by a foreign ruler were taken as a symbolic acknowledgment of suzerainty of a type that could be traced back to the Sage Emperors. The fact that normally an envoy came, not the ruler himself, perhaps symbolized the ruler's partial separation from the imperial political order as well as his involvement in it.

A naive political realist might think this emphasis on appearances irrational, a distraction from or a cover for the realities of power. But it certainly had realistic functions when the foreigners involved shared much of the ideology and ceremonial idiom of China, as did the Vietnamese, the Ryukyuans, and above all the Koreans, or were at least ambivalently impressed by the power and splendor of Peking, as were many Inner Asians. But, when the foreigners involved shared none of this, and did not think of themselves as presenting tribute, the appearances did not convey to them the impressions they were supposed to convey. Officials were not always much disturbed, however, by indications that foreigners were unimpressed and insincere participants in the tribute system. We have noted late Ming cynicism about sham embassies from Muslim Inner Asia. In these four case studies, it is clear that Board of Ceremonies officials knew that in the "memorial" brought by Saldanha the King of Portugal did not call himself "Your Minister" and that European monarchs did not use Ch'ing reign periods or seals granted by Ch'ing emperors in their "memorials." Nor did the embassy routine embody any exceptional effort to make an impression on the foreigners; if it had, envoys probably would have been received in more spectacular, less matter-of-fact ceremonies.

Officials could afford to be a little casual about the impression they were making on foreigners because the embassy routine was not oriented primarily to using ceremony, documents, and regulation to make the right impression on the foreigners. Rather, its primary audiences were the metropolitan officials who attended ordinary audiences, the officials and common people who saw the embassy party pass by in the capital and elsewhere, and the posterity that would read the records of the dynasty and pass historical judgment on it. This primacy of domestic audiences fits nicely with the primacy of domestic politics mentioned above. The embassy routine was superbly adapted to the preservation of appearances before these audiences. The web of bureaucratic regulation insured that an embassy's contact with individual Chinese would be limited, that it would be only temporarily in the capital, that it would follow much of the pattern of events and ceremonies laid down for previous embassies, that it would travel under close supervision beneath banners announcing it as a tribute embassy, and that, when it appeared before the assembled metropolitan officials in the courtyard before the T'ai-ho Hall, it would be announced as a tribute embassy and the ambassador would perform the ceremonies appropriate to that position. Because of restrictions on an embassy's private contacts and the length of its stay, it was not likely that any private record would survive to challenge an official "re-translation" of a non-submissive "memorial."

Some later European ambassadors would recognize the Chinese effort to keep up tributary appearances and struggle against it; the four Ambassadors studied here reconciled themselves to these appearances in various ways. The Macaenses of the Pereira de Faria embassy had been doing what the Ch'ing authorities told them for many years. Manoel de Saldanha was determined to defend the honor of his King, but the Jesuits convinced him that the change in the banner of the embassy meant that it was not being received as a tribute embassy. By the time he was ordered to prostrate himself in the court before the T'ai-ho

Hall, he probably was too sick to care. In any case, both Portuguese embassies were under close Jesuit guidance. The Jesuits were very anxious for them to succeed and had enough experience of the court to know that no embassy could be received without conforming to the tribute ceremonies. The Dutch, as seen in the records of the decision to send the Van Hoorn embassy, understood that their embassy would be taken to be a tribute embassy, and did not worry about the appearances of the system as long as they understood that embassies treated in this fashion included many from princes who in reality were powerful and independent. It was a perfect fit; the Dutch worried about realities but not appearances, while the Ch'ing rulers insisted on preserving appearances even when the realities did not match.

The long-run consequences of the preservation of appearances in these four embassies were considerable. Coming in a period of relative realism in policy-making and of moderation of traditional Sinocentrism, they were one factor in the failure of the Ch'ing state to emerge from these decades of relative openness with any new policies, attitudes, or institutions that would help them to pay attention to and deal with the changing realities of Sino-Western relations over the next 150 years.

As I suggested in my brief discussions of Sung and Ming diplomacy, the focus on ceremony, and thus on appearances, insured that a dangerous reliance on illusion would be a persistent failing of Chinese foreign policy. The early K'ang-hsi court, however, showed a substantial ability to overcome inherited illusions and to respond realistically to changing circumstances. The Manchus came to power in China after a century in which the tribute embassy had received little bureaucratic attention, had been subject to the most cynical manipulation by foreigners, and many aspects of foreign relations had been managed without reference to it. They were themselves the descendants of the most adept manipulators of the Ming system, and as foreigners ruling China scarcely could be expected to swallow whole traditional ethnocentrism. In dealing with substantial threats from the Khalkha and the Russians they were in no way hampered by Ming precedents and illusions. They maintained traditional forms for the reception of embassies but did not rely on manipulation of prestige and gifts to manage the danger; they began the great advance of Ch'ing military power into Inner Asia, and even negotiated a treaty with the Russians in 1689. The court also was showing some capacity to recognize that Chinese convictions of cultural and political superiority were not fully applicable in relation to maritime Europeans. The culture that had produced the learned Jesuits and the astronomy, mathematics, and music that were studied at court might not equal the Chinese in subtlety and good sense, but clearly it was a high culture completely autonomous from the Chinese in its origins and development. The fact that Europeans were not required to use Ch'ing dates and seals on their "tribute memorials" may indicate some awareness that they were not quite the same as other tributaries.[2]

The main reason why this relative realism and openness led to no substantial innovations in policies toward maritime Europeans was that the Dutch, the Portuguese, and the English simply were not as important to the Ch'ing policy-makers as were the Mongols and the Russians. The Dutch were only temporarily useful as naval allies, and proved dangerous and intractable even during those years. Between 1660 and 1690, Sino-European trade was in a slump, producing little revenue for the court and its agents, and no great pressure from foreigners for increased access. (The changes in control and taxation of maritime trade in the 1680s were primarily directed toward Chinese shipping, and were applied to foreign ships in Chinese ports almost as an afterthought.) In these circumstances, it is not surprising that there were no institutional innovations, and that the court's appreciation of aspects of European culture had almost no impact on traditional diplomatic forms.

But the traditional forms did not just go unchallenged; they were revived and incorporated in the Ch'ing regulations and precedents as a result of the early embassies from the Dutch and the Portuguese. This was not primarily a result of Ch'ing policy. In striking contrast to the early Ming, the early Ch'ing rulers made no comprehensive effort to bring all known foreigners into tributary relations with their court.

The reasons for European participation were varied, transient, anecdotal. The Dutch sought relations with the Ch'ing in the 1650s because they were not satisfied with their trade on Taiwan. They were told they would have to pay tribute before they were allowed to trade, and in 1656 suggestions that they be allowed to trade in non-embassy years were rejected; but these were decisions after uncertainty and debate, not foregone conclusions, and the latter was linked to efforts to cut off trade with the Cheng regime. In view of hostilities in 1665 and decisions at the end of 1666, the insistence that no trade would be allowed in 1666 unless an embassy was sent seems less a demand for an embassy than the leaving of one small opening in a relation in which the court, if not the provincial officials, had lost interest. Although the Ch'ing authorities insisted in 1684 that no more Dutch trade would be allowed until an embassy was sent, it is doubtful if this exclusion could have been sustained very long in an order that asked no questions about foreign traders' tributary status. The Portuguese were confirmed in possession of Macao before they sent an embassy. They never were told they had to send one, but sent two on their own initiative.

Thus, these embassies were sent for a variety of reasons, in which European initiatives and European conceptions of what an embassy ought to accomplish were very important and the results deeply ironic. The Dutch quite rightly saw the first years of a new dynasty as a period of relative plasticity of institutions, when old precedents might be ignored and new ones established. They hoped to establish more favorable precedents for their trade by negotiation with the court, but the Ch'ing officials, partly because warfare with the Cheng regime led them to forbid non-tributary maritime trade, responded by reviving old Ming precedents, applying them in full rigor to the Dutch, and almost immediately losing interest in them. We have seen this in the failure to establish a fixed route and period for Portuguese embassies, in the willingness to tolerate Dutch failures to send embassies on schedule as long as there was any prospect of naval cooperation with them, in the acceptance of the rather transparent sham of the Pereira de Faria embassy, in the severing of the trade-tribute link in the 1680s, and in K'ang-hsi's remark after the Paats embassy that embassies were too costly and it would be just as well if none came. This double movement of accidental revival and prompt boredom and neglect left at the core of Ch'ing attitudes and policies toward maritime Europe a blurred patch of institutions and precedents that were neither actively enforced nor explicitly rejected, that effectively blocked any re-thinking of the Sinocentric core of the Chinese diplomatic tradition. This would not necessarily have prevented realistic and effective policy-making that tacitly ignored that core; this is what happened in Sino-Russian relations, despite occasional Russian participation in the tribute system. But, until about 1745, there were no serious political problems involving maritime Europeans to attract the attention of Ch'ing statesmen; even then, the difficulties centered on missionaries and their converts, and the maritime traders attracted little sustained attention. In the absence of such attention and realistic policy-making, the illusions sustained by records of early embassies probably were more important for the long-run development of Ch'ing attitudes toward Europeans than they would have been otherwise. Moreover, these illusions were occasionally reinforced by further European involvement in the tribute system.

It is important to remember that over a hundred years elapsed between the Paats embassy and the next appearance in Peking of an envoy from a major European maritime power.[3] The embassies of that century all were focused on the problems of the Catholic missions. The papal legations of Maillard de Tournon and Mezzabarba were rarely if ever called tribute embassies and were not treated according to the strict embassy regimen, although kowtows by the legates and condescending favors from the emperor maintained the fundamentals of imperial ceremonial superiority. Their stays in Peking were longer, their opportunities for negotiations with the emperor greater, than they would have been within the embassy routine. The third papal legation, of Plaskowitz and Ildefonso, was called a tribute embassy, recorded as one in the *Statutes and Precedents,* and handled largely according to embassy protocol.

The Portuguese embassy of Alexandre Metello de Sousa e Menezes, received in Peking in 1727, was sent by the King of

Portugal in an effort to persuade the Yung-cheng Emperor to modify or abandon his anti-Christian policies. When the Ambassador left Canton for Peking he was allowed to bring a suite of over 40 and received from the Ch'ing authorities an advance of 1,000 taels toward his travel expenses—both departures from the routine for earlier embassies. Imperial officials acknowledged to the Portuguese that the embassy could not be compared with "the common embassies of ordinary tribute-bearers," but it was recorded in the Ch'ing records as a tribute embassy. The Ambassador objected to every instance in which he was referred to as "paying tribute," and insisted on being called an envoy "presenting congratulations." He had with him copies of some of the documents of the Saldanha embassy, and on this point probably was following the precedent established by that embassy, but, as we have seen, this change in terminology in no way placed the embassy outside the tribute tradition. In Peking he insisted on being allowed to present his credentials in person to the Emperor; this was granted, and duly recorded in the regulations as a variation in embassy routine that could be allowed if the ambassador had orders from his sovereign to insist on it. He kowtowed repeatedly and without hesitation. On the advice of the Peking Jesuits, he did not bring up the missionary problem in his meetings with the Emperor, and left with nothing more than the faint hope that he had gained a little good will for his monarch and his faith.

Another Portuguese ambassador, Francisco de Assis Pacheco de Sampaio, was received in Peking in the summer of 1753. He reported that he received assurances that he was not being received as a tributary ambassador, but in fact he was so recorded in the Ch'ing records; since this was a first embassy to the Ch'ien-lung Emperor, it probably was labeled an "embassy presenting congratulations" as its predecessor had been. The Emperor received the Ambassador at the Ch'ien-ch'ing Gate in the Palace, received the royal letter from his hands, later was present at one of the banquets for him, had his portrait painted by Castiglione and came to see it, gave the Ambassador a jade good-luck scepter (*ju-i*), and invited him to enjoy the dragon boat races at the Summer Palace. The Ambassador performed "the customary ceremonies" on all occasions. These were extraordinary gestures of favor, but they were largely personal gestures of the Emperor, unrelated to the statutes and precedents of the tribute system. Neither this embassy nor any other European embassy between Paats and Titsingh was received before the full capital bureaucracy at an ordinary audience in the courtyard before the T'ai-ho Hall; this suggests to me that formal receptions of European ambassadors were slipping into a kind of inner-palace twilight, only marginally related to the forms and bureaucratic regulations of the tribute system. Pacheco de Sampaio had been sent to seek some relief from new repressions of Catholicism in the empire and especially in Macao, but, like his predecessor, decided not to mention the missions to the Emperor.

Far more important and far more paradoxical in its effects on Ch'ing policy and Sino-Western relations was a last burst of embassy activity, beginning over thirty years after the embassy of Pacheco de Sampaio. In the 1780s, both changes at Canton and changes in European politics and ideas gave new impetus to efforts to open formal state-to-state relations with China, bypassing obstructions and vested interests at Canton and bringing the messages of science, progress, and unrestricted trade directly to the Imperial Court.

An English embassy was sent in 1787, but the Ambassador died on the way out. The effort was renewed in 1793 with the dispatch of a large and imposing embassy under the distinguished Lord Macartney. It carried an impressive range of samples of English manufactures and exhibits of the latest scientific advances. Macartney was to seek changes in trading conditions at Canton, cession of a depot, and permission to establish an English resident minister in Peking. Once arrived, Macartney turned a blind eye to the banners that proclaimed him a tribute ambassador, but rejected all urgings that he practice the kowtow. He suggested that he would be willing to perform before the Emperor any ceremony a Ch'ing official of equivalent rank would perform before a portrait of George III he had brought with him. The Ch'ing officials rejected this proposal, and finally he was allowed simply to kneel on one knee before the Emperor. But this was at an audience in a great tent in the grounds of the imperial summer estate at Jehol. The breach of protocol was witnessed by many high court officials, but it was not in Peking, not before the assembled metropolitan bureaucracy. It is very unlikely that it would have been permitted in the courtyard before the T'ai-ho Hall. Ch'ing official records give no hint that anything unusual had taken place. The requests were rejected on the ground that there was no precedent for them. An English envoy resident at the capital would violate the empire's "fundamental system"; that is, it would violate an unstated but obvious basic principle of the tribute system, that ambassadors came for short ceremonial visits, not for long residence and substantial negotiation. This request and Macartney's refusal to perform the kowtow were the first cases in which the Ch'ing tribute system, rather than restriction on coastal trade or missionary activity, became the focus of Sino-European controversy.

These challenges, coming when a culturally conservative autocracy was facing major problems of corruption and rebellion, stimulated completely negative and defensive responses. Coastal officials were ordered that, if any Europeans wished to send or carry letters to the Imperial Court, the officials were to refuse to accept them if they contained requests for any new privileges or anything contrary to precedent. But even such a troublesome embassy represented a revival of European participation in the tribute system, and an illusory incorporation in it of the most powerful of the Canton trading peoples. Thus, the embassy at once provoked defensive reactions and revived old illusions. The illusions were further strengthened in 1794 by the return to Peking of the first European tributaries, the Dutch.

Some of the impetus for the Dutch embassy came from Ch'ing officials at Canton, who seem to have been worried about the complaints about them brought by the Macartney embassy and anxious to find some other group of foreigners who would send a less troublesome embassy. They told the Dutch chief at Canton that the Portuguese and the English were planning to send embassies to congratulate the Emperor on the sixtieth anniversary of his reign, and surely the Dutch would not want to be left out of this important occasion. He reported this to his superiors at Batavia, who promptly organized an embassy, naming Isaac Titsingh Ambassador. Arriving in Canton in September 1794, the Ambassador found that no other Europeans had sent an embassy, and that there never had been much chance that they would do so. The Canton officials made sure that the embassy would be beneficial to them and useless to the Dutch by demanding a promise from Titsingh that he would not make any requests or complaints in Peking, but was going simply to congratulate the Emperor; he gave the promise.

The embassy then was hurried off to Peking so that it would be present for the jubilee ceremonies on the Chinese New Year, making the cold, miserable winter journey in only 49 days. The imperial court, apparently delighted by the arrival of such a tractable and undemanding European embassy, treated it with many signs of favor. The ambassador performed the kowtow on many occasions. He was present at the great New Year's banquet given for tributary envoys in the Pao-ho-Hall, a formal occasion primarily for Inner Asian and Korean tributaries in which no previous European envoy had participated. The presence of tribute envoys at a New Year banquet was one of the longest continuities in the history of Chinese foreign relations, stretching back to the early years of the Later Han. Seeing the Dutch there, the old Emperor and his court must have found it easy to put the Macartney unpleasantness behind them, to forget about the tedious financial disputes at Canton, the growing opium imports, and all the other strange people from the West who traded at Canton but had never paid tribute, and conclude that the traditional order symbolized by the tribute system still was a living reality even in relations with Europeans.

The Titsingh embassy was the last occasion on which a European appeared before the Son of Heaven within the tribute system. The Macartney confrontation had convinced Europeans that they could avoid the kowtow, now fixed upon as the system's most demeaning feature, if they were firm enough, and at the same time had made the Ch'ing authorities determined never again to give a European embassy any leeway to attempt any deviations from prescribed ceremony or to make any troublesome requests. The Russian Ambassador Golovkin was turned away in 1806 after he refused to perform the kowtow before a tablet representing the Emperor, probably in token of gratitude for a banquet, at Urga. In 1816, the Amherst embassy, in absurd violation of the usually leisurely schedule of a tribute embassy, was insistently summoned to audience on the very day of its arrival in Peking, and sent away when the Ambassador refused to appear; probably the officials in charge had hoped that in this way the Ambassador could be hustled into performing the kowtow before he had a chance to commence a long discussion about appropriate ceremonies, such as that which finally led to Macartney's violation of tribute-embassy protocol.

Ch'ing policy toward maritime Europeans drifted toward the great confrontation of the nineteenth century isolationist, preoccupied with issues of ceremonial and documentary precedence, seemingly unable to focus on the realities of the intrusion into its world of great powers that did not accept or even tolerate Chinese practices in foreign relations. Isolationism, ceremonialism, and a focus on appearances rather than on realities outside China also were characteristic of the institutions and regulations of the tribute system. The ceremonial core of that system assaulted by Macartney in the kowtow controversy and the request for a resident minister were even more vehemently defended after 1842. Thus it is not hard to see why, especially when looking back from the nineteenth century, it has seemed to make such good sense to refer to the whole pattern of isolationist, appearance-obsessed, Sinocentric foreign policy as a "tribute system." But to do so is to blur the difference between Ming foreign relations, which were based on a comprehensive tribute system, and the much more differentiated Ch'ing system. We have seen that, even in the decades of the apparent revival of a comprehensive tribute system, from about 1654 to 1684, embassies were not central to the pattern of relations with the Dutch and the Portuguese, and that eighteenth-century policy toward Europeans was made with almost no reference to the embassy institution. The comprehensive-tribute-system explanation also ignores the fact that Ch'ing statecraft scholars discussed foreign relations not in terms of tribute but in terms of defense.

The Ch'ing writers' focus on defense suggests one way to reintegrate what we know about the tribute system in the less comprehensive and more exact sense used in this book with a general interpretation of the Chinese diplomatic tradition. The institutions of the tribute embassy were focused on one form of defense, the defense of the appearances of the ceremonial supremacy of the Son of Heaven in the capital. Thus, in it, defensiveness and ritual order, the core themes of my description of the Chinese diplomatic tradition in *Pepper, Guns, and Parleys,* were unusually closely intertwined. But this was not the only or the most important form of defensiveness in Ch'ing policies toward Europeans. They also sought to defend the empire against cultural and political subversion by missionaries and their converts and against the dangers to public order of foreign ships and sailors along the coast and in Chinese ports. The relations among these forms of defensiveness were complex and changing, but only in the Macartney embassy were the tribute system and the particular defensiveness it embodied central to them.

The whole complex of defensive policies made excellent realistic sense for late imperial China, with its solitary eminence in East Asia, its impressive but rather thin and passive bureaucratic control. These policies were administered by statesmen sometimes capable of highly realistic and systematic action. Why, then, did they end in the self-destructive clinging to illusions and forms that characterized the Ch'ing court's relations with Europeans from 1839 at least to the imperial audience of 1873 and in many ways until the defeat of the Boxers? These failings would have developed even if there never had been a single maritime European embassy received within the tribute system. (This is not hard to imagine, and except for the Macartney embassy it would not have had much effect on the course of these relations before 1800.) Chinese foreign policy frequently had slipped from appearances into illusions ever since the Sung. The tribute system would have been an important part of Ch'ing foreign relations no matter what the Europeans did, reinforcing illusions of Chinese supremacy, tendencies to cling to their ceremonial expressions, and a general defensive orientation, all of which contributed to the great conflict with the Europeans.

But it was European involvement in the tribute system—absent-minded, unsystematic, shaped by strange provincial contexts and by European illusions, never accepting the ideas expressed in the system—which, if it was not essential to the survival of Ch'ing illusions, gave them firm institutional form, enshrined them in dynastic records and regulations, and made it much more difficult for even the most clear-headed Ch'ing statecraft scholar to see past them and to focus on the great change toward which Chinese foreign relations were drifting.

Liang T'ing-nan was a distinguished scholar who lived most of his life near Canton, participated in the compilation of the *Compendium on Kwangtung Maritime Defense,* and was the chief compiler of the *Gazetteer of the Kwangtung Maritime Customs.* As a well-known local expert on the management of foreign trade and on maritime defense, he was consulted by Lin Tse-hsu as soon as Lin arrived in Canton. In the *Gazetteer,* completed in 1838, three chüan out of thirty are devoted to the precedents and regulations for the tributaries coming via Kwangtung. But *after* the Opium War, in addition to publishing works on England and America drawing on foreign sources, he published an *Account of Countries That Pay Tribute via Kwangtung,* that is, of the Siamese and the maritime Europeans as tributaries. Its preface lists the various countries that sent tribute to the Ch'ing; Korea, Ryukyu, Vietnam, Japan (which never was a Ch'ing tributary). "Beyond these there were the countries from the most distant corners of the Western seas, who never before had been in touch [with China; now] there was none that did not turn toward civilization and make sincere offerings, fearing that they would be late to present tribute. . . . Among those who came to pay tribute via Kwangtung were Siam, Holland, and from the Western Ocean, Italy, Portugal, and even England. All their envoys and ships came at the proper times. . . . There were regulations for welcoming them, officials to accompany them . . . [These documents] enable us to discern the principles of unifying all and leaving none outside, the far-reaching institutions of the Sage Virtue cherishing those from afar."[4]

Appendixes
Notes
Bibliography
Glossary
Index

appendix a

Brief Account of the Journey Made to the Court of Peking by Lord Manoel de Saldanha, Ambassador Extraordinary of the King of Portugal to the Emperor of China and Tartary (1667-1670)

Written by Father Francisco Pimentel, S.J.

Published by permission of the Houghton Library, Harvard University, the Biblioteca da Ajuda, Lisbon, and Professor C. R. Boxer.

Note: There are two manuscript versions of this text, Palha #1 and Ajuda 49-IV-62, fol. 715-732. Both apparently are copies of a lost original. C.R. Boxer and J.M. Braga published a generally accurate annotated transcription of the latter in 1942.[1] *The Palha text includes one substantial passage and a number of phrases that are not in the Ajuda text, and has plausible alternative readings for a good many words and phrases that do not make very good sense in the latter. In preparing this translation I have assumed that some readers will want to follow it or check points in the Boxer-Braga transcription. Numbers in parentheses correspond to the beginning of each page of that text. Footnotes include (in addition to comments on assertions of fact and explanations of names and terms) transcriptions of the Portuguese text on which my translation is based where it diverges from the Boxer-Braga transcription. Most of these divergences result from my adoption of the Palha text; the very few that result from a reading of the Ajuda text different from Boxer and Braga are labeled "Ajuda." My translation has been checked by Bobby Chamberlain.*

Before beginning the account of this journey, it seems to me necessary to give brief notice to the affairs that led to the request for this embassy, which may not be well known to all who read this document. First, they must know that in 1658 Nanking, the Southern Court, which is another Niniveh, was attacked by the long-haired Chinese, who are natives of the Province of Fukien, also called Chincheo, whom the Castilians of Manila call Sangleys,[2] who inhabit not only that province, which is on the mainland, but also various islands that are near it, and now also possess the island of Formosa, which they took from the Dutch. Among these Chincheos those who live on the mainland obey the Tartar

Emperor, like the rest [of the Chinese], but those of the islands who are numerous and the strongest and most warlike[3] of this people, never subjected themselves to him, nor did they wish to cut their hair, and took as their leader the son of I-kuan.[4] [This I-kuan], a baptized Christian who was brought up in this City of Macao in a little tavern where his parents sold things to eat, by his industry acquired a little ship, and in it became a pirate in the seas of China, with such success that if he had been in Europe he would have inspired fear in its greatest lords, (12) for, in addition to the infinite host of his fellow-countrymen that followed him, he came to have with him on the seas more than 10,000 junks, the majority war-junks, which are like our frigates, although different in design, and mount 40 guns. Many of them are made with such artifice that no cannon ball can penetrate them, and others remain afloat and sail on as before even though penetrated by a thousand balls, because, it is said, they are founded on beams linked to each other in the form of a raft, an invention that frightened the Dutch.

With this power he attacked Nanking, and, after a siege, had almost taken it and subjected all the surrounding territory. But at this time fell his birthday, which the Chinese are accustomed to celebrate with infinite drunkenness. All were stretched out drunk on the field, and, treachery being so natural to this people, some Chincheos went over to the forces in the City, and informed them of the condition of their enemies. Soon a little more than 3,000 Tartar cavalry came out, and began slaying the drunken ones at will, meeting little or no resistance. Those who escaped took to the water and fled on their junks.

The Emperor, being informed of this success, was not content with the victory on land, and tried to follow it up at sea. He sent from the court many people, among them many Tartars his favorites and relatives, who embarked and set out in search of I-kuan,[5] who was enraged by the failure of the siege. And, since the Tartars were brought up in the mountains and have among their people no one with any knack or talent for the sea, those who were there were cowardly so that I-kuan attacked them and killed them all, not a one escaping.

Feeling this misfortune, and considering the power of the enemy, and the peril in which they had put Nanking, the Emperor assembled his Council and asked how it was possible that I-kuan, being lord of so few lands, could have the capital to sustain so many people and such great fleets. Among the matters that were pointed out there was one, which by reasons of the injuries which it brought to so many millions of souls could only have come from Hell by the mouth of some demon; and it was that I-kuan prospered by trade and commerce with China itself, from which he obtained all the drugs and merchandise that he took to Japan, and from Japan he brought the silver with which he sustained his power. Nothing more was necessary to move the barbarous Emperor to take one of the most cruel and tyrannical resolutions the world has seen; for he ordered that all cities, towns, and villages which were near the ocean withdraw four leagues inland, in five provinces, each one of which is larger than all of Portugal. It is not to be believed, nor can it be reduced to numbers, the infinite multitude of people that perished because of this decree, nor the loss the Emperor himself (13) suffered in his revenues, for just in the Salt River, which is near Macao, he lost 18 million taels in revenue every year, each tael being a third more than a pataca.[6]

This prohibition of trade [*sic,* for the evacuation order] reached Macao in 1662, the year when the Emperor commanded that in all his empire not a single plank should put out to sea, nor should there be any commerce with people from abroad, on pain of death. And since the inhabitants of this City have no lands of their own, nor even a handful of earth on which to fall dead, to take trade away from them was the same as taking away their lives. But in the first year they sustained themselves with the capital they had accumulated, hoping that the Emperor would see that I-kuan flourished as before without the trade with China,[7] knowing the damage he caused to his subjects and incomes by this barbarous decree, realizing that with it he did no damage to his enemies, but to himself and his subjects, would lift this prohibition. But, being a tyrant, and fearing to lose the Empire he had usurped, he continued with great rigor the observance of his black decree, so that the people of Macao resolved in November of 1667[8] to request urgently of the Viceroy of India, the Lord Count of São Vicente, João Nunes da Cunha, that he send an ambassador in the name of the King Our Lord, so that going to the court, and speaking with the Emperor, he would represent to him the miserable condition of this City, and receive from him some dispensation[9] of trade with which they could sustain themselves, for only along this road was there a glimmer of hope of relief from so many evils. The Lord Viceroy did not find anyone who would dare to take upon himself such an arduous enterprise, except Lord Manoel de Saldanha, who was moved more by zeal for the Faith, which in these parts depends principally on the conservation of this City, and for the service of the King Our Lord, and by no other interest. Scarcely recovered from a serious illness, he embarked in Goa, and, having in his voyage passed through horrible dangers from which his escape was clearly miraculous, arrived ill at this City on the fourth of August and soon embarked ill for the City of Canton, capital of this province, where its governors detained him for over two years, for reasons I will recount below, until finally an order arrived from the Emperor which commanded that, without any further exchange of correspondence, he depart soon for the court.

The Lord Ambassador embarked at the City of Canton on 4 January 1670, with the state and splendor that suited his person and (14) authority. He wore crimson satin trimmed with silver, the ornamental slash of his hat, his sword-belt, shoulder-belt, and his collar all of silver. [On the river boat] the floor of his room was entirely covered with carpets, which are much esteemed in China; [the room also contained] a canopy of red damask with fringes of gold, beneath which went the letter and portrait of the King Our Lord; two buffets with panels of the same damask, the fronts of red satin fringed with gold; eight chairs of red velvet with gilded fringes and ornamental nails, six low chairs with the same velvet and fringe; and the sides and doorway of the room were ornamented with curtains of damask fringed with gold and silk. The multitude that came out to see him was infinite. The Viceroy of Canton, and the *t'i-tu,* Generalissimo of the Sea,[10] whom we met on the way, had the windows of their halls opened so that they too could enjoy such a spendid spectacle.

And, since the court had ordered that the number of people who were to accompany [the Ambassador] should be limited, not to exceed 22, not counting the Lord Ambassador and 3 of his officials, it was not possible, despite many efforts, to secure the agreement of the mandarins in Canton that the servants should not be counted in this number. The Lord Ambassador departed accompanied by 12 Portuguese men, all the rest being his servants or slaves.[11]

The principal vessel went in front, always flying the standard and arms of the King Our Lord, and in this way they crossed all this great empire until they reached Peking. It also carried a yellow banner in the Chinese style, with letters which the Fathers had given us, which said: "This is the Ambassador of the King of Portugal, who comes to congratulate the Emperor of China." When these people consented that the word *chin-kung,* which means tributary, should not be placed on this banner, this was the greatest victory over their pride that the great prudence and tact of the Lord Ambassador could achieve, for, as they said, this broke the custom which had continued for more than 2,000 years, that no Ambassador enter[12] without the title of tributary.

In all the towns and cities where he arrived, the highest dignitaries who governed them always visited him with gifts. At the customs houses, of which there are ten along this route, where all vessels customarily are searched, because of the goods they carry, those of the Lord Ambassador (15) always were treated with great respect, they not daring to enter them. These rivers by which one navigates to the court are always full of innumerable vessels which carry the tribute of rice; for those of the Emperor number 10,000, in addition to those of the merchants, which are infinite. Thus, even the great mandarins have great difficulty navigating, because the river[13] is so full. But with the Lord Ambassador they used terms of singular honor and favor; in Huai-an and in Lin-ch'ing [in each of which] one of the highest officers of the empire was in charge, he sent a mandarin of his household with authority and power to order the vessels sent away, and the way cleared. Around this depot[14] there are many gates that serve to hold the waters to make sufficient depth for these vessels to navigate. They are as big as our caravels or pataxos; some of the larger ones carry more than 4,000 piculs, leaving free the lodgings of the mandarin who goes on them. Each picul contains 140 arrateis.

These Chinese are politic to such a high degree, and so extremely exacting in insisting on the respect and preëminence due to their ranks, that, when two mandarins meet along this route, the inferior quickly has his towrope[15] lowered, so that the superior may pass. [Which is superior] and his rank, is quickly known by the banners I mentioned above, and others[16] in gold which they carry carved on tablets on the bow superstructures of their vessels, so that all can easily see who it is, and give way, as all who met us gave way to the Lord Ambassador. But so that we would miss no experience in this journey, while we were still in the province of Peking, the Lord Ambassador met a brother of the Petty King of Fukien who, because he had not lowered his towrope, advanced on the same side of the river where our towrope went. A message was sent to require him to give way and pass on the other side. He replied that we could pass[17] more easily to the other side, because we went with the water, and he was not about to move away from there; so that he found his honor so much offended that he finally resolved that the dispute should be referred to the court so that there it could be determined who should give way; this is customary in such cases. While this was going on, something happened which all judged to be a special work of the Divine Providence, very nearly a miracle. Suddenly, furious storms of wind and rain came up, although up to that time the weather had been clear and serene. The hawsers[18] of some junks were broken, so that our junks, and that of the Lord Ambassador, ended up against the other side of the Canal. Five or six of the junks the toll-collector [comptidor] had with him also broke loose, and only his own[19] did not cross, since on it only the hawser broke which they customarily attach to the stern in order not to block the Canal, and impede the way by which other junks pass, and the bow hawser held fast, so that the wind moved the junk around but it did not lose its place. (16) Thus Heaven settled the dispute in this way, putting one party on one side of the Canal, and the other on the other. Night came, and the next day each went his way in peace.

So that those who wish to know of the authority and respect with which the Lord Ambassador was treated throughout this empire may avoid the tedium of reading many identical cases, I will describe only one event that occurred when he entered the province of Peking. It came about as follows. As was customary, a petty mandarin, a Chinese in the service of the Lord Ambassador, went ahead to present to the official of the town the *k'anho,*[20] which is an order of the Emperor which orders that silver be given him for the expenses and the towrope coolies for a distance of 80 li, which is about 6 of our leagues, a little more or less. Our mandarin had such a hot discussion with the mandarin of the town about the collection of the money and the sending of the coolies that they laid hands on each other, and then came to blows, so that he wounded or caused him to be wounded, so that blood flowed; which is not done among this people;[21] so he had him arrested, and brought from his house to the boat of the Lord Ambassador, where he slept that night, bound hand and foot, in the sight of his people, who, although they came to help and called out, did not dare to take him from the hands of, nor use force against, him who had brought him there. Only one who knows the power and authority of any official in China can understand how much respect was shown to the Lord Ambassador on this occasion. The next morning the Lord Ambassador sent him [the local official] back to his house, recommending that in the future he be more careful and diligent in executing the orders of the Emperor, and, so that he would not be entirely without consolation, had given to him some European goods as gifts, so that he returned to his house more contented than he came, and we went on our way.

In all the cities and towns where he arrived,[22] the officials have a royal hostel where they are received, and all expenses are paid by the Emperor, more or less in conformity to the rank of the person. But the Lord Ambassador never wanted to enter any one of these hostels, but stayed on the boats. And in truth they are so comfortable that there is no need to go ashore, for the boat has in the middle two cabins and a parlor, all very capacious, with their buffets and chairs. They are painted and gilded with a thousand decorations and fine carvings, amply provided with windows, which in bad weather are closed with wooden shutters, and in good, mild weather are opened, and still the windows are covered with gauze, fastened like curtains but stronger and firmer, so that the official can see all that goes on outside without being seen by anyone. Toward the stern and bow are more cabins for the servants of the official. These are over another deck, on top of lower houses, in some of which the sailors live and in others the trumpeters. Around this housing, all around the boat goes a corridor 5 palms [about one meter] wide by which service and communication are maintained throughout the boat. These boats are larger or smaller according to the capacity of the rivers, but the smaller are as comfortable for the official as the larger although in a smaller space. From Nan-ch'ang, capital of Kiangsi, to the court the boats are so large that they look like ships. On the rivers nothing is lacking necessary for life or for pleasure. Indeed it can be said with truth that one Chinese city should be counted as two, one on water and one on land, for there are so many boats, and the number of people who work on them is so enormous, that they cannot be counted. Some are large, some small, of a thousand types and styles. Many sell everything necessary to sustain life. Others, to avoid the effort of hawking their wares, carry certain signs and little flags which proclaim what they are selling. I also found in this a custom worthy of much attention, which is that, when they place some goods on sale, they raise up a cross, so that in China a cross is a sign of a proclamation which is exhibited to anyone who may want to buy. This is a use of the Holy Cross worthy of much consideration by those who know it. Everywhere there is an abundance of fish, so that it seems that it could be no greater either in quantity or in quality. From the lake that begins as one sails to An-ch'ing, which is a city more than 40 leagues before one reaches Nanking, is taken that famous fish which in Portugal we call *sothe,* which here is of extraordinary size, for some weigh more than 9 arrobas [c. 135 kg]; we bought a cross-section measuring more than a span [c. 20 cm] of the length of the fish, and it weighed 70 catties [over 40 kg] which are 100 of our arrateis. The price is very low, for each catty cost a cash and one half, which would be 2 arrateis for less than a vintem in our money. Although very large, they have no bones except for some very soft cartilages and stem-like structures; on the outside they are not altogether covered with scales, but have some large ones that look like strong plates; but the meat is so fat that it resembles beef. Thus, the boats being so comfortable and the rivers so full of everything necessary for life and pleasure, the officials come and go from one province to another with their whole families so leisurely that they seem more to be on a pleasure outing than making a journey, and thus they do not resent the hardships of travel.[23]

The Lord Ambassador passed the greater part of it [the journey] in grave illness and distress, for in the beginning of March he was stricken by an erysipelas [acute skin inflammation] in one leg, with unbearable pain, which then turned into large wounds, which persisted until he arrived at court. At the end of April, he was overcome by diarrhea so importunate and continual that it made him very weak. In this condition, he reached the end of the journey, 4 leagues from Peking, where our Fathers who serve there came to visit him, and to give him information[24] on what was to come. Three days after arriving at this place the Lord Ambassador left for the court, on the last day of June, in the company of a mandarin whom the tribunal of the Lipu [Board of Ceremonies] sent to expedite all that was needed for the journey. On the same day in the evening (at the beginning of the night) the Lord Ambassador arrived at the court where he was lodged (17) by singular honor and favor not in the ordinary houses[25] for ambassadors, but in the palace of a great man which had recently been confiscated. On the next day, he went to the tribunal of the Lipu to present the letter and portrait of the King Our Lord; the Presidents received him standing, with much courtesy. The presents also were turned over, which were not very splendid, and much inferior to what they expected; so that we all judged that one of the greatest favors the Emperor did to us was to accept them, excepting two cotton cloths, which, because they had painted on them with little propriety two obscene figures, the Presidents said might not appear before the Emperor, and for this reason they gave them back to us. Hypocritical pretense of these heathens who, having within themselves the same turpitude, in public wish to shine with that modesty and virtue that they destroy in secret! Evident proof that they know the good and do not follow it, but content themselves with appearing to have it. But there is still something worthy of imitation in this action, for at least they have respect in the eyes of men, which many of those who should have it lose, even in their own eyes. One should not wonder at the insufficiency of the presents, for they had been assembled in this City, so exhausted and consumed by the more than eight years preceding.[26]

After the Lord Ambassador presented them, the Ministry of Ceremonies had several questions asked of him, of which these were the principal: Why was the word *ch'en,* which means "vassal", missing in the letter from the King Our Lord? The Lord Ambassador gave the reply which the Fathers had indicated to him before he went to court, that is, that in Europe it was not the style or custom when kings wrote to one another that some called themselves vassals of others. This was the question we had had fixed in our imaginations on the whole journey, for it placed everything in jeopardy, since we were determined not to speak as they wished in this matter. The Fathers in Canton armed us against this question with the letters they gave us on the yellow banner of which I spoke above, for, when the Chinese wrote after the name of the Lord Ambassador *chin-kung,* which are two letters, of which the first, the *chin,* means "enter", and the *kung* signifies "tribute," the Fathers changed the second letter *kung* to *ho,* which means "congratulations," and wrote *chin-ho,* meaning "enter to give congratulations." These letters were on the banner from Macao to the court, and from the court to Macao, with such outlandish novelty, and amazement of all, that only God could have caused it. They also asked if the King Our Lord would send another ambassador to China, and the Lord Ambassador replied that he could not know what the King would do in the future, and thus could not respond either yes or no. After this answer, the order of the Emperor arrived that they not (18) ask any more questions, and we returned giving a thousand thanks to God because he had delivered us with honor from such ignorant questioners.

After the Lord Ambassador reached the house, his diarrhea gripped him again, and, when the Emperor learned of his condition, the care and solicitude with which he had him cared for were remarkable; he multiplied so much his gifts and favors, his special messages, that that whole great court was thrown into astonishment. He ordered the two best doctors and the mandarins of his palace and our Fathers that they should go to visit [the Ambassador] every morning and evening, and bring him news immediately on the same day of how they found him. One cannot appreciate these favors unless one knows that the Emperors of China are treated and respected not as men, but as gods, considering themselves and calling themselves Sons of Heaven. Never since ancient times, in more than two thousand years, is there record that any emperor gave to anyone the honors that were given to the Lord Ambassador. This mad presumption which seems to want to surpass human limits is based on, in addition to his lack of the Faith, the astonishing greatness of his empire; truly, if many of the Kings of Europe could see with their own eyes the vastness of his territories, the greatness of his cities, the richness of its commerce, the infinite number of its revenues, the teeming world of his court and the magnificence of his palace, and finally the majesty of his state, and the variety of the great emblems and dignities with which he appears in public on his throne on the 1st, 5th, 15th, and 25th of every month, it is certain that they would feel very small before this great monarch and would come to believe the truth of what has been written, that any one of the fifteen provinces of this empire is more rich, more vast, and more populous than many of the kingdoms of Europe.

The Lord Ambassador was in bed until the 29th of July. On the 30th, he went with all his people to the Board of Ceremonies, where the ceremonies that customarily are carried out before the Emperor are learned, which are to kneel three times, returning to standing each time, and upon each kneeling to lower the head three times until the brim of the hat touches the ground,[27] and if, by accident, the hat should fall during these ceremonies, that is a great calamity, for this people considers the uncovering of the head a very boorish and discourteous act. There the mandarins informed the Lord Ambassador that he might not wear a sword when he appeared before the Emperor, for this was a law that was kept inviolate, not even his own brother being exempted. The mandarins gave him this admonition in such a good manner and with such courtesy that they seemed to be asking a favor rather than informing him of the law.

This had its origin a little over two years before when great men of the court with the chief of the Regents, Patrocum (19), plotted to kill the Emperor with their short swords when they went to knock their heads before him on the days of the month I mentioned above. Nevertheless, the Lord Ambassador replied with such vehemence that many thought it was excessive. But so great was the respect in which he was held and the desire to please him that they consulted three times on this matter, but never dared dispense with a law that was obeyed so rigourously by all the great men of the empire; they feared the eyes of the Prince, which necessarily would witness the dispensation that he had not granted. This law applies only in the presence of the Emperor, for, when the Lord Ambassador returned to the Palace to receive the presents for the King Our Lord, we all wore swords. On the 31st, he went to the Palace for the first time, dressed in black Persian camlet, a novelty among the Chinese in that it was not made of silk, and therefore much esteemed, trimmed in silver, the ornamental slash of the hat, sword-belt, shoulder belt, and collar all of silver. The chair in which he went, a little smaller than a palanquin, had no cover, for [the cover] was of crimson cloth of gold, and its curtains of cloth of gold. The mandarins said that such a cover might not go to the Palace, for those colors were proper to the Emperor, and only he could wear them; if it had another cover, even if it were richer, full of pearls and diamonds, it could be taken as long as it was not in those colors. We soon saw the truth of what they said, for all that belongs to the Emperor was the said colors, even the roof-tiles being glazed in yellow and red. The closest relatives of the Emperor on the paternal side wear yellow sashes; those more distant or on the maternal side, red.

Having arrived at the Palace, the Lord Ambassador and we performed the customary ceremony to the Emperor in the open court, where it also was performed by over 5,000 officials, for only they enter there. These ceremonies are performed to the sound of various instruments which are played in a hall which is[28] on that same open court in front of the royal Throne Hall.

The signal for the beginning of these ceremonies is eight crackings of some whips, like coachmen's but incomparably larger, and so heavy that they never rise from the floor, and make as much noise as eight pistol shots. And, so that no one shall make a mistake and the ceremonies not be disturbed, there is an official standing,[29] who calls out in a loud voice when one should kneel, knock one's head, rise, and so on, so that all will perform uniformly and at the same pace. The Royal Hall is very large, all gilded, with ornate carving in gold, blue, and red. It has 70 columns in two ranks, which divide it into 3 naves, like our old churches. One goes up to it from the open court by 5 flights of steps of a very white stone like marble, beautifully worked. Partway up the steps are 2 landings with railings of the same stone, of painstaking and perfect workmanship. On the spaces between the flights of steps are various (20) incense burners of gilded bronze, with the capacity of a barrel. On the last of these landings where the steps end, the Royal Hall runs lengthwise, giving a view much like that which the Church of the Hospital gives on the Rossio in Lisbon, except that here the Hall runs lengthwise, while there only the facade [faces the square]. This Hall has three portals that open onto the terrace; in the front of the central, principal one is the royal throne, almost 3 fathoms high, all of carved wood in all perfection of the art. The place where the Emperor sits has neither canopy nor back, but is like a table from whose sides issue two serpents, superstitious arms of this empire, so arranged that with their coils they form two beautiful ornaments between which the Emperor appeared seated, and, if many of us did not deceive ourselves, he had his legs crossed on the table. It has no back, for from the open court where we performed the ceremonies we saw him seated from the waist up, and behind his shoulders the light and air that entered from the portal on the other side that corresponds to the principal one. I write this in such detail because it is very difficult to conceive of a building by means of a written account, as is seen in the variety that is depicted for us of those referred to in Holy Scripture.

When these ceremonies were completed, the Emperor had us summoned to his Royal Hall, where only those of highest rank, the Princes and Colaos, attend, near to his throne, and spoke to him with striking affability and respect, had tea given to him, and, when he spoke of him among his own people, did not call him the great vassal of the King of Portugal. On the same morning, when the Lord Ambassador already had returned from the Palace or I should say when he was on his way back,[30] [the Emperor] had him summoned again and spoke to him in private, causing universal astonishment at such an unaccustomed favor, in a veranda near the women's quarters, where no man enters except some eunuchs. And, because he wanted to show more favor, he had Father Ferdinand Verbiest serve as interpreter, and Father Luis Buglio also was present. After talking to him, he had distributed to the Lord Ambassador 10 pieces of silk, to the three officers [of the embassy] 6 [each], to the others 4; all were damasks, velvets, or satins, and some with gold backgrounds. This was a particular favor, in addition to the ordinary, for later when [the Emperor] ordered the delivery of the presents to the King Our Lord, the Ambassador received 100 taels in silver and 25 pieces of silk, some with gold background; the three officers 12, and 50 taels of silver; the others 20 taels, and 6 pieces of silk.

(21) The Emperor did us another favor worthy of record, by which he showed his esteem for the Portuguese, when the President of the Lipu presented him a memorial of the presents for the King Our Lord and for our people, which were equal to those customarily given to other kings and ambassadors, [31] the Emperor reproved him saying that these were very few, and that the Portuguese should not be thought of like the Dutch and the Siamese, but were people from very far away, from the Country of the Great West, and that this was the first time they came to his lands, and for this reason the presents should be increased. And he asked if the Lord Ambassador and the people were given good food and wine; for, as long as we were at the court, all expenses were on his account, and with such generosity that we were given five times more than was customarily given to other ambassadors, and every day 3 arrobas of snow, for it was very warm. All this honor and esteem that the Emperor showed we owed to the Fathers, for they had given the Emperor advance

notice, with ample information both orally and in writing about Europe and about Portugal, and about the Lord Ambassador, giving him particular information as to who he was, and of the ancient and illustrious family from which he descended, so that the great men of the Palace treated him as a Count, for the Fathers had introduced him with his dignity. Outside [the court] people venerated him even more, for they gave him the title of Prince.

But so great was the passion of our malcontents that, seeing this with their own eyes, and the singular benevolence with which the Emperor treated the Fathers, the gifts and favors he bestowed on them, sending to their house the deer he killed on the hunt, and the fish he caught in the river of his Palace and the fruits of his gardens, which being connected with royalty they sent on to the Lord Ambassador; our malcontents said they came from the marketplace or bazaar, and added that it was worth nothing to the Emperor; at the same time that they knew very well how much he favored and respected them, and that the Fathers entered not only to the sixth open court as was allowed to the Lord Ambassador and us as a special honor, but farther into the women's quarters, where no one may go, and there he frequently sent them food on his own golden plates, a favor never bestowed on anyone in China. Now see what credit they deserve in these matters and in those concerning the Fathers, who, because they joined with the Lord Ambassador in order to serve him, and guide him in all his negotiations, incurred the disaffection of these malcontents, as all those who went with us to the court know, saw, and experienced. They are so abandoned (22) to their passions that no evidence will suffice to make them confess the truth. But, so that it would be impossible to deny the special protection with which God aids the Fathers, it was fitting that, while we were still at court, the Emperor gave to Father Ferdinand Verbiest the same dignity that had been held by Father John Adam, and made him President of the Tribunal of Mathematics with a new favor, that he presides alone without a colleague, contrary to the custom of this and all other tribunals in which two men always preside, one Chinese and one Tartar. In our presence the mandarins congratulated him on the new promotion; our malcontents cannot deny it no matter how much they may wish to do so. But the Father accepted only the care and work of this office, but not the title, nor the state of such a great mandarin, because the Fathers believed that this was suitable for the greater benefit to Christianity and the service of God.

After the presents were received there quickly followed the banquets, of which there were three, in the Tartar style. At the first, in the name of the Emperor the mandarins attended at table in the same robes and ornamentation with which they attend the Emperor in public, and which are only for this use. The other two banquets were, as is the custom, in the name of the Lipu, and in each one of the two Presidents of this Board was always in attendance. No one should imagine that these banquets have any resemblance to ours in ostentation, civility, and cleanliness; quite the contrary. In the first place, the tables are raised only two palms from the floor, and they use neither tablecloth, nor napkin, nor knife, nor fork, and all the meats that are offered are so underdone and badly cooked that they seem raw, so that one cannot eat without losing one's composure, grasping with the hands and tearing off with the teeth, like some glutton or sheepdog. The meats were of sheep, cow, pig, horse, donkey, chicken, goose, and duck, without any other benefit of the cook's skill than a boiling. For this reason none of us ate anything, and it seemed that we had been invited only to look on. Also offered were many fruits, fresh and dry: chestnuts, walnuts, hazelnuts, grapes, apples, peaches, very large and good, of almost all the kinds there are in Portugal. One or two disasters happened to me at these banquets, to which I cannot refer without asking the good and special indulgence of the pious reader. At the first two banquets, they put before me the head of a sheep, with two horns so large that they frightened me. I don't know how they found me, or by what sign they knew me, in order not to mistake me and to present me with these points two days in a row, (23) for I did not sit in the same place. The head was so little cleaned that by its wool I knew that the sheep had been black. One should not be astonished that I refer to such a base matter in such clear words, for I believe it behooves me to oppose the zeal and attention with which some people so exaggerate the civility of the Chinese that they prefer it to our Europe, and want to place it above our heads. Now they should see what kind of understanding one can have who is so devoid of good breeding as not to recognize the incivility, baseness, and discourtesy of [serving] a head so ill prepared; for this was in China, and even at its court, and in its palace. I acknowledge in the Chinese much politeness, grandeur, and richness, but in all these things there are intolerable basenesses. They are very polite, but put before you on the table a dish like that; they have many grand things in the empire and its cities, but their houses are like barnyards; many are rich, but there are among them infinite numbers of poor, who have nothing of their own save their poverty; they are very clean, but the plates which they served on the three days at the three banquets were never washed, and, when I asked why they were not washed, they replied that that silver was issued by weight and returned by weight, and if they washed it many times it would be diminished and they would have to make up what was lacking, and for that reason they never washed it. With this I conclude this argument, wanting everyone to know it, even if it cost me some complaint. Before beginning the banquets the President of the Board of Ceremonies performed the ceremonies mentioned above to the Emperor, facing to the north where his palace is; we and the Lord Ambassador also performed them with the President; for this is a custom which everyone observes at this court.

Finally, the Lord Ambassador returned to the Palace and spoke with the Emperor a second time with the same favors and in the same places as the first time, with this difference; it being a courtesy and ceremony which we all were ordered to observe, and saw observed by everyone, to cross running the open way directly in front of the Emperor, on our second visit we were ordered to stand in the middle of it. Some considered this a great favor, and so it was in truth, but it also is certain that it was done because of the great curiosity of the women, who were watching what was happening from behind incompletely closed doors behind the Emperor, and if we did not stay still in the middle they would not be able to see us, or in order for them to see us it would be necessary to open the door so far that they would be seen too. It was rumored in the Palace that the Portuguese were all noblemen; and, if the reader wants to know the truth, there were few there who merited that title, save by their offices, for these Tartars are for the most part so ugly, so enormously fat, so loathsome in aspect, that among them our people might indeed appear as noblemen, as the dim-sighted (24) are as lynxes in the land of the blind. Our Kaffirs also played their role there, for the Emperor had them come close to him, and ordered them to open their mouths, and show their teeth, and asked what they ate. [We replied] that they ate very well everything that men eat, and drink much better. Soon he ordered two mandarins to wash the Kaffirs on one shoulder, for he suspected that that black color was artificial. The two doctors who came to treat the Lord Ambassador had given foundation to this suspicion, since they also treated a slave of the coolie caste, a native of Bassein [in India] ; in his illness, although he did not turn white, he became less black, and somewhat pale, for the natives of India are not as black and dark as those of Mozambique. They soon went to tell the Emperor that it seemed to them that, since this black color turned to white in an illness, it was not natural. From this arose the curiosity to have them washed, which had no other result than that they lost some of their stench, which was transferred to the washing-jar.[32]

At this second visit, the Lord Ambassador described the predicaments of Macao, to which [the Emperor] replied that he already knew everything, since the Fathers had referred to all this many times. He did not present the memorial on the maritime trade of Macao, as those of Macao wished him to do, for, when he took counsel with the Fathers, they answered that it would not be suitable, for the reasons I will mention farther on.[33] There the Emperor asked many other things: If there were falcons[34] in Portugal, and if our King had some; by this he signified that he wished to have some, but, so that it would not appear that he who lacks nothing needed something, he did not dare request them. He also asked the Fathers about water spaniels and retrievers, and if they had the abilities that were told of them; for this Emperor was much inclined to the hunt. He also asked if we had emus or little lap dogs, no doubt because he had been told that we had two with us, but ours were more suitable for the laps of dwarfs,[35] quite unlike those wandering around in the Palace, very large and heavy, but still very much esteemed. Finally, he asked about weapons with blades, buff waistcoats, and so on, and all this was to signify that these were the things he wished and desired; for of all the other things we had brought him he already had a surplus. I learned from the Lord Ambassador that he was considering having these things sent to him, and others which would be esteemed because they had never been seen there. Our long swords that bend back on themselves were much esteemed, and on our visits we soon were asked to show our swords. I refer to these trifles, so that if at some time there is another journey like this one it may be known what may give pleasure there. To these princes it is as they say to us in their language "Kity," which means "a new thing," and the Fathers add that, if the new things were (25) of great value, it would be much better, but even if not it would suffice for it to be new to be much esteemed, and the ship that brought them would be admitted without difficulty, and without paying anything.

The Lord Ambassador endured much discomfort in all these visits because of weakness resulting from his illness. It grieved all of us to see him go along more and more wretchedly, with the color and expression of one dug up from the grave. The Palace, in which neither horse nor palanquin may enter, contains great expanses, and, since [the Emperor] spoke to the Lord Ambassador in the sixth courtyard, and they are so vast that every one is equal to or larger than the Terreiro do Passo in Lisbon, it was necessary to go a long way on foot. The mandarins told us that the Dutch and Siamese neither saw nor spoke to the Emperor, but only reached the fourth courtyard, where they were told to knock their heads before the walls. But, so that the reader of this may form some conception of the size of this Palace, I will mention here some things from which it can be deduced, for one could never finish with a description of it, and I did not see all of it, nor even half the interior part. The area it occupies is that of a great city. It has within it various streams suitable for boats like our pinnaces. Two of these streams divide two of the courtyards through which we passed, passing between masonry walls, in such a form, that they represent a bow, and the bridges, which are very elaborate and perfect, seem to be arrows in the bows. More than 16,000 souls are employed within the Palace. It has four gates in its walls, facing the four principal winds. When the long-haired Chinese ruled, at each one of these gates there were on duty 5,000 guards and 5 elephants, and all had the lodgings within the Palace and within its walls. But, now that the Tartars rule, they do not have so many soldiers in the Palace, but all the people of their guard, who are more than 200,000 men, live in what is called the Inner City, for the court [capital] is divided into three cities, each one with its walls, gates, etc. The first is the Imperial Palace; the Tartars occupy the second, called the Inner City; and the Chinese live in the third. In the Inner City where we were, and where the Fathers also live, when we went to visit them we walked more than a league.

The climate, location, and dwellings of this capital are very bad, for, although it has many fruits, it has other intolerable inconveniences. In the winter the cold is unbearable; therefore all the beds are over furnaces.[36] [The winter] begins so early that this year on (26) the 9th of November snow fell to a depth of one span, or better one yard, and soon the rivers froze.[37] In summer the heat is excessive, and a greater torment is the dust, of which there is so much and fine that, when we went out on the street, our hair and beards looked like those of millers covered with flour. Bad water; innumerable bugs that get in one's clothing at night, and while we were there they bit many of us. The flies are infinite and importunate, and the mosquitoes worse. Everything is very expensive. The streets are not paved, as it is said they were in the past, for the Tartar ordered the removal of the paving stones, because of the horses, for in China they don't know what horseshoes are, and this is the reason why there is so much dust, and when it rains unbearable mud. One who hears of the grandeur of this capital will conceive of something like Lisbon, Rome, or Paris, but, so that he will not be deceived, I warn him that, if he entered it, he would think he was entering one of the poorest villages of Portugal, for the houses are all built very low, for they may not exceed the height of the wall of the Palace, and so badly made, that the walls for the most part are of mud, or plastered wattle, and very few with any bricks or any view outside, and all China is like this. In the houses of the great and the rich, all their grandeur consists of large courtyards surrounded by rooms, like a church, much painted inside but offering very little comfort, for they give no protection against the heat in the summer, nor against the cold in the winter, being open like covered porches, with lattice windows covered with paper much finer than ours. One can see how much protection and shelter this offers from the ravages of the weather, severe as they are in Peking. I should be given some credence in this matter, for in two and one half years I reached five provinces, all the way to the court, and I also was in Chao-ch'ing, capital of the province of Kwangsi, which is the sixth [I have been in].[38] I was in many tribunals and mansions of viceroys, and at the court I was lodged with the Lord Ambassador in the mansions of one of the four Governors who governed this Empire during the minority of the present Emperor. The Fathers were not pleased that I formed such a low opinion of Chinese buildings,[39] and, in order to dissuade me, said that I had not seen everything, nor had I been in the house of a certain mandarin, or in such and such a city, which were works which could appear [without disgrace] in any part of the world; to which I replied that those they mentioned might have all the perfection of art, as they said, but those I had seen had been as I write here. All this meanness of the buildings arises from the circumstance that, being rich, (27) their first care is to conceal their wealth so that the officials will not work up some swindle (as is their custom) and take it all from them. Someone may want to know, China having so many provinces with such beautiful and mild climates, why the court is in the worst of all, that of Peking? The reason is nothing other than the convenience of being close to the frontiers, 7 leagues from the Great Wall, in order to encourage the defenses against a Tartar invasion, and so that the presence of the Emperor may keep a check on the perfidy of the Chinese against the loyalty and unity of the Tartars among themselves; which is so well known of the Chinese, that they have a saying that one Chinese has one hundred hearts, and one hundred Tartars only one.

On the 20th of August, the Lord Ambassador was given permission to visit the church and house of the Fathers, where he spent the whole day, and left a large contribution for the Holy Liturgy. On the 21st he went to the tribunal of the Lipu to receive the reply of the Emperor to the King Our Lord, a most singular honor which had never before been bestowed in China, for the Emperors do no customarily send replies to any King,[40] as the Dutch, Siamese, Tonkinese, and Cochin Chinese could testify in our times, for all these came with embassies but none brought a reply. That to our King was not only the first, but as the Fathers (to whom the Lord Ambassador sent it to be read) said to me, it was notably modest, its expression quite different from their usual vainglory and presumption. The Lord Ambassador left the court on the afternoon of the same day, in his chair carried by eight coolies, the chair covered with crimson cloth of gold, fringed with the same, and on the corners above,[41] and at the peak two poles of beaten silver, the curtains of gauze (*lo*) shot with gold. There is no other like it in or out of the court. The coolies all wore red robes. The Ambassador departed dressed in blue satin ornamented with gold, the ornamental slash of his hat, his sword-belt, shoulder-belt, and his collar all of gold. The multitude of people that came out to see him was not to be believed, some calling to others to come see this spectacle, calling in loud voices that they should come to see the Petty King of Portugal; they gave him this title because it is the highest among the Chinese. Also many Tartar ladies came out to enjoy this sight, accompanying us, some in covered carts like coaches, others on horses, on which they place themselves like men, seeming to be so in dress, manner, and impudence, even in the smoothness of their faces, for the Tartars keep themselves as hairless as women, and do not want to have any hair or beards, and if by mischance any hair should appear they quickly pluck it out.[42] The Emperor sent in company with the Lord Ambassador a mandarin of the Lipu to expedite matters on the route to Macao; the Fathers also accompanied us until the Lord (28) Ambassador was embarked on the royal serpent boats, which were seven,[43] all painted inside and out, even the masts, so that when they are under way they make such a beautiful sight that they seem like bouquets. A little after we had embarked, two kaffirs escaped from us and went back to the court, but, as soon as the Emperor learned of it, he quickly sent two mandarins to come in person and turn them over to us, and ask that we not punish them; what made this all the more to be esteemed is that the mandarins carried Kam ho [*k'an-ho,* travel documents] to come to Macao with the Kaffirs if necessary.

The Lord Ambassador began his voyage to Macao on August 27, and was regaining his strength so rapidly that we all persuaded ourselves that he had health and life enough for long years to come. But, in the province of Nanking, after passing the Yellow River, on a Saturday, October 18, he had such a severe attack that he was soon left muddled, and his judgment disturbed; the doctors came, and, although they did everything for him they knew how to do, nothing helped; that which we regretted most was that he was not able to confess himself, nor could he take the Holy Viaticum. But we have great faith in God that he received His mercy, because of the singular devotion with which he always venerated this highest Mystery, and moreover he received absolution and extreme unction.

The Lord Ambassador died in Huai-an, like many other places all flooded by the inundation of the Yellow River, on October 21, leaving us all afflicted with cares for eternity, and memories of his having suffered much in the service of the King, and especially of Macao. But, as a calamity rarely comes alone, but brings many with it, so also the death of the Lord Ambassador brought with it the disorders of the secretary Bento Pereira de Faria; for, while the Lord Ambassador lay in agony, which lasted almost the whole day, three or four hours before he died this man came in, accompanied by four manservants of the embassy, showing so little consideration with shouts and disorderly voices that they not only disturbed him, but perturbed me in the ministry in which I was then engaged to help the Lord Ambassador, who was still conscious, at that critical moment (29), and as some Holy Water was being put on him he trembled. This was an action that gravely scandalized everyone, not only by its lack of respect and decorum toward his superior, but also by its lack of compassion and Christian piety in an occasion and circumstances when it is most proper and most expected.

With these disordered shouts, over which the Chinese were amazed, looking at each other, he began to declare that now he was the Ambassador, and asked for the keys, not only those that belonged to the embassy but also those of the Lord Ambassador, and all his baggage; he annulled the patent that the Lord Ambassador had given to Andre Coelho Vieyra in which he made him his lieutenant in charge of the whole embassy and his replacement after his death, and removed him from this office which he had occupied for nearly two years while the Lord Ambassador was alive, saying that the Lord Ambassador did not have the power to give this office to anyone except to himself, to whom it belonged because he was Secretary. I say that Andre Coelho Vieyra must be given much credit for his behavior on this occasion, for he protested the compulsion that was imposed on him and on his tenure in office, not overlooking any measure which he was obliged to take, and much more for the moderation and prudence with which he repressed the impulses natural[44] to a military man, which he is, and did not resort to force and violence when he saw himself despoiled of his dignity by one who had no right to it; for, if he had used force or resorted to arms, it is certain that the Emperor would have known of it, as he knew of everything else, and would have considered the Portuguese insolent and rebellious people, to the great detriment of Macao, of which they have always been afraid; and, if there had been dead or wounded, the embassy party would have ended by going to the tribunal with chains around their necks to hear their sentence; those guilty of homicide are decapitated, and others are beaten in public.[45] Let us consider now how important it was that this succession[46] should go to a prudent man! Bento Pereira de Faria considered none of these arguments, and to me who proposed to him those that can be found in the papers of the embassy he said that I might be a good theologian in the Company [the Society of Jesus] but as an archivist there was no one who was better placed than he, who knew very well what he was doing, and who would give a full account of everything in due time.

All that I have referred to up to now were faults that could be passed over, for after all they did not get out from among us [Portuguese], or at least not off the boats; but I cannot omit mention of that which was done thereafter, for it was public, and it is proper that all should know the truth. As soon as he had himself recognized as the successor Ambassador he took from the bedside of the Lord (30) Ambassador the letter from the Emperor and ordered that it should go on his boat,[47] which went in the third place. Moreover he took the yellow flag of the same Lord with the Chinese letters that say "This is the Ambassador who after giving congratulations to the Emperor returns to his King," and put it on his boat, which he had go in front as the flagship, that of the Lord Ambassador coming after it, causing great discredit and scorn. And, because he did not dare move the standard of the King Our Lord or his royal arms, he left them on the boat of the deceased, a greater affront than if he had moved them; for now the royal arms went behind Bento Pereira, following the back of one of his subjects when they never came after any king in the world. The Chinese, of whom we expect less because of their great haughtiness, let them go hoisted in front throughout the Empire, and did not put them behind their yellow banner, as Bento Pereira now did.

It is not easy to describe how scandalized this nation [the Chinese] were by this, or the variety of reactions they expressed; for some were indignant, while others laughed at us, and all asked, "This mandarin, who you said was so great, now after his death when he should be more honored, who made him petty? And why do you take from his place in front, and his flag? Our King [Emperor] did not do this, nor that of Portugal, who cannot yet know what has happened in China." All of these irregularities were pointed out to Bento Pereira before he resolved to do them, and he was warned in front of everyone what offense he was giving to our King, to his Ambassador, and to all our nation, and all this was done on paper. But he paid so little attention to these warnings that later he came up with a new one, in the opinion of many worse than the previous, and this was that making an inventory of the goods of the Lord Ambassador, in front of those who were present he opened the chest of the deceased and took out of it all the papers, among which there must have been highly secret matters, for he [Saldanha] never entrusted the key to this to anyone, but until death always carried it tied to the waistband of his drawers. We did not fail to protest to Bento Pereira on behalf of the Ambassador—or I should say the Viceroy—that he should not touch those papers, for there were all the secrets of the embassy, and other affairs of the court, and of Macao, which only concerned the state. There can be no end to an accounting of the evil he did, for, even if there were no papers there except letters of his household and family, by what title or in what way do these secrets belong to Bento Pereira? Moreover, many knew that there were papers of great importance there. I can affirm as one of the Lord Ambassador's most intimate companions for more than two years that, if the state of glory and blessedness which we trust that he is enjoying from God were capable of grief and resentment,[48] the Lord Ambassador would feel them greatly toward Bento Pereira's opening his chest and taking his papers, for, in addition to the usual reasons for (31) grievance this action contains, the Lord Ambassador had other very particular reasons to resent it. Also much to be resented was the poorness, or, to put it better, the baseness, of the coffin which he had bought to hold the corpse of the deceased; for, besides being of pine, a very ordinary wood here, it cost, as he said, not more than 10 taels, or God knows how much, it being the custom in this empire, even among people of middling means and rank, to buy for one's burial a coffin worth at least 60 taels. But, in addition to costing so little, it turned out so badly that, after eight days, as his corpse decomposed into blood it could not hold it, and it ran out at two places. Bento Pereira could have used more properly on honors for the Lord Ambassador the silver he added to the pay of the soldiers with whose support he took over, for the Lord Ambassador had given them 3 patacas per month but now he gave them 7½.[49] It seems that he considered it legal and necessary that those who took his side in this weighty matter should have some profit from it.

As soon as the Lord Ambassador died, the Governor of Huai-an and the mandarin who accompanied us sent couriers to the court to inform the Emperor. It is certain that the Emperor was much grieved by this death, for he clearly had desired good health for the Lord Ambassador, and in no case wanted him to leave the court until after the eighth moon, at the beginning of October, for he said he did not want him to go if he was not completely well. But he did give permission, because the Lord Ambassador sent to have him informed that if he could not leave by the end of August at the latest he would not be able to arrive in Macao during the [north] monsoon and would be compelled to wait there another year. After news of his death arrived at the court, Father Gabriel de Magalhães, Superior of that residence, wrote to me that the Emperor showed great grief, and expressed it immoderately to Father Ferdinand Verbiest when on the next day he went to present to him a memorial on mathematics. The Emperor immediately sent an edict with an order to do honor to him in whatever city the courier should reach us; when this order reached us, we were already in the city of Canton, where the following honors were carried out. First, he ordered the *hsien cheng-t'ang* [district magistrate], who is a great mandarin, to come knock his head to the corpse of the deceased, in the name of the same Emperor. He came with many flutes, shawms, clappers, and drums; five tables were set up, on which were pigs, goats, many kinds of cakes, sweets, wine, and (32) some dried fruits, all as offerings, and many sticks of incense which were burned before the coffin. The mandarin knocked his head three times, and three times poured wine into a cup of gilded silver, and poured it out on the earth.

He also brought various castles and towers made of paper, of gold, and of silver, and among them came a paper from the Emperor, written in very large letters which contained an elegy of the Lord Ambassador, which was read aloud before all the people, and, because I could not have it in my hands, I asked one of the Chinese who heard it to relate to me what it said. He replied that the elegy contained in summary this general notion: "You, Manoel de Saldanha, Ambassador of the King of Portugal, obliged me very much, for you crossed the sea and underwent much travail to come and bring congratulations to me. I, being grateful, unable to do any other favor and not knowing where you are, offer you these things, and dedicate them to your honor and memory. Accept them with[50] my good wishes." The Emperor also ordered that, if we wished that his body remain in China, we should select a place for his tomb. In the afternoon of the same day the Petty King of Canton sent a mandarin of his household to knock his head on his behalf, with the same pomp and ostentation. The next day the Tsung-tu and Fu-yuan, that is, the First Viceroy of his provinces and the Second Governor of Canton, did the same. No one should imagine that these ceremonies, offerings, and burnings of incense are in any way sacrifices, or acts of religion, for bronzes have no part in them, and no divinity is invoked, but all this activity is merely civil and politic, instituted solely to honor the dead, whom the Chinese, more than any other nation, respect and venerate.

A few days after these ceremonies concluded we left for Macao, where we arrived on Tuesday of Holy Week, five days after we left Canton. We supposed that Bento Pereira, being now in the lands of the King Our Lord, and under his fortresses and artillery, would correct the disarray he had made since he took over, for the Captain-General of Macao, João Borges da Silva, to whom we wrote from Nanking [possibly meaning the province] had ordered him by a letter, which was given to him as soon as we entered this province [of Kwangtung], that he should come in the same form as the Lord Ambassador came from Peking, but he ignored the orders of the Captain-General, and entered Macao in the same manner as always; he went in front[51] (33) with the flag he took from the Lord Ambassador, and the standard of the King Our Lord with the corpse of the deceased behind on another boat, and what astonished us was that no one hindered him nor even found it strange.

After he arrived at Macao, he began to put it about that the Fathers in China had set out to make the embassy a failure for this City, for they advised the Lord Ambassador not to present the memorial of which I spoke above, but he did not relate the reasons to the Fathers mentioned why it did not behoove the same City to present such a memorial; he spread about these and so many other things so contrary to truth, to all that we found on this journey and saw with our own eyes to be the truth, which he denies having seen or experienced, in three and one half years, that the little respect he has for everyone[52] causes astonishment. For it is certain and evident that all the honor, favor, and esteem that the Lord Ambassador found in China is owed to the Fathers who serve there, both in Canton and at the court. This will not be difficult to believe for one who will consider dispassionately that, although the Lord Ambassador had the quality and nobility we all know, he was no better known in China than the Dutch or Siamese, and they only came to know him through what the Fathers said of him and of the greatness of our King. The presents which he took from the City were more likely to lose respect for him than to gain it; for we knew that, when the President of the Lipu saw them for the first time, he clearly scorned them, and taking in his hand the string of coral which came for the Emperor posed this question and line of reasoning: "This coral doesn't come from Portugal?" (They consider that everything that comes from Europe comes from Portugal.) Our people replied to him that it did; then he replied "Why do the Dutch bring from your land such big corals, and you such small ones?" The replies that our people gave were so mixed up and confused that I don't remember them, but it can be seen what they must have been. One should not forget, however, that of the son of the same President, who was in attendance there and was a young man with better manners than his father, to whom he said, "Sir, we do not esteem this embassy for what it brings us, for they do not come here to make us rich, but because such a great personage sent by his King comes here from so far away to offer congratulations to our Emperor. They do not come to buy or sell or request titles or factories like the Dutch and the Siamese." With this reply the President was silenced. It can well be seen that the public esteem the embassy had in this empire was entirely based on the information provided by the Fathers both about the King Our Lord and about the person and nobility of the Lord Ambassador, for, as soon as he arrived in China, they at once were asked who he was and if they knew him.

(34) The Fathers performed other services for the King and his embassy on this occasion, from which all can see how far from the truth is the account given by Bento Pereira. It is an old custom of this nation that no one can discuss matters with any high official unless he first knows what the matter is, and to what it is related; an extremely politic practice, in order never to have to answer rashly in an important matter. In conformity with this custom, the Emperor ordered that the letter from the King Our Lord be opened, and that the Lord Ambassador be asked if he came to discuss any business[53] of Macao, or for what did he come? He replied that he did not come on any business of Macao, for, although the King of Portugal knew of the condition it was in, he only sent them to give congratulations to the new Emperor of the empire which he had conquered. The Lord Ambassador replied in this manner for two reasons; first, because the Tsung-tu [Governor-General] who hanged himself had informed the court that the Lord Ambassador came by order of Macao to discuss its business, and was not sent by the King; the second because, if it was said that it came about Macao, that would make it all the more certain that it would not go to court, nor be received. At three different times the mandarins tried to open the letter from our King, and never opened it, for they wanted the Lord Ambassador to open it with his own hands. This being a matter that would arouse so many suspicions, he replied that he could not open the letter from his King, and, if the mandarins had an order from the Emperor, there was the letter, and they should open it. The third time they came intending to open it, but, finding that the pouch was fastened shut with the seal of the Royal Arms, they did not dare to open it, but again informed the court, and, since it is more than 500 leagues from Canton, in the coming and going of all these messages two and a half years were used up. The malcontents blame the Lord Ambassador for this detention, for they say he should have opened the letter when the mandarins asked him, giving no heed to the irregularity that

would have been involved, and the reasons that he had; for this might have been a device to test him, to see in how much respect he held the letter, from which they would infer if it was from the King. He also is blamed for not visiting the Petty King [Shang K'o-hsi], not seeing the expenses that would result from such a visit, and the obligation it would impose on him to visit the Tsung-tu [Governor-General], Fu-yuan [Governor], and Pu-cheng-shih [Provincial Treasurer]; who are the higher dignitaries [after Shang], and, if he had not visited them, he would have made them his enemies, and, if he visited them, it would have led to insupportable expenditures, for soon after the visits come the banquets and theatricals, at which the guests customarily give to the performers at least 60 taels, according to their rank, and soon invite [their hosts] to another banquet. And Macao's support was so limited that in Canton the Lord Ambassador twice had his table silver melted down. (35) In addition, he would have lost the respect the Emperor had for him when he [the Emperor] learned that he [Saldanha] did not want to see anyone in China before he had seen him [the Emperor].[54]

Finally, a firm order came from the court that the letter was to be opened and translated from Portuguese into Chinese. For this translation Fathers Feliciano Pacheco, the Vice-Provincial, and Francesco Brancati were summoned to the Tsung-tu, and to them the Lord Ambassador had secretly given a transcript, so that they could take out or put in what seemed suitable to them. They translated it so that it lost nothing of the majesty of our King, nor left in anything offensive to Chinese haughtiness. This transcript they quickly sent to the Fathers at the court so that they would agree with those in Canton, when the Emperor ordered them to translate it. The letter contained some words the Fathers did not dare to remove without the approval of the Lord Ambassador, saying to the Emperor that it was hoped that His Majesty would give entire credit to Manoel de Saldanha in all the business he would discuss. The Fathers told the Lord Ambassador that, if those words were translated, it was certain that he would not go to court before he had said what these business matters were, and, if they concerned Macao, he would never be admitted. The Lord Ambassador, after much consideration, ordered them not to translate them. If this translation had been done by hands that were not so zealous for the King and for the good of Macao as were those of the Fathers, this embassy never would have been respected or admitted. Truly, the Dutch experienced the truth of this in this same year in relation to a memorial they sent to court, which was not translated with the greatest[55] good will, and the translation of this letter was the reason why they were ordered expelled from Canton, with so much abuse and bad treatment that even we who have so few ties of obligation to them, were caused to feel sadness and compassion, for they were Europeans. They complain that the Fathers could have done a great deal of good for them at no cost to themselves, by removing an equivocation contained in the memorial; they have reason for the complaint, and the Fathers also have reason not to help them, and they know well why.[56] The Fathers performed another service for the King Our Lord and his Ambassador, which he [the Ambassador] did not cease to enlarge upon and admire. In the visiting papers and memorials that the Lord Ambassador presented, his scribe and interpreter called him *chin-kung,* which means tributary, and the cover of these memorials was dark blue. These memorials usually were delivered by Antonio Aranha de Barros, to whom His Majesty surely would give great favors if he knew how much he exerted himself in his service on this embassy. He took one of these memorials to give to a certain mandarin, and with it in his hand went to visit the Fathers, who, seeing the memorial, and the cover that it bore, had him tell the Lord Ambassador that they wished to warn His Lordship that his scribe and jurubassa (who are the interpreters) were discrediting him, for they called him (36) tributary and enclosed it in a blue cover, which was the ceremony and style of vile and base people, and, there being in China no other privileged group besides the literati, they never use such a cover, and it was even less suitable for His Lordship. They also admonished him to use his personal name, Manoel de Saldanha, instead of the word *Ambassador,* which is the title of an office which the Chinese customarily translate *chin-kung,* or tributary, for they have no other word for it in their language. With this advice the memorials soon went with red covers, which are those of the great people, and the Ambassador put his personal name on them, and this impropriety was corrected.

As soon as the Lord Ambassador was ordered to go to court, the Fathers instructed him concerning the flags and insignia he should take to give proper ostentation to his grandeur in the Chinese fashion. And, because it was of great importance for the honor of the King Our Lord and of the entire Portuguese nation, the Chinese letters *chin-kung* to which I referred above, which mean "one who enters to present tribute," were changed; they removed that letter *kung* and put in its place *ho* which means congratulations; the two together, *chin-ho,* mean to enter to give congratulations. This was without doubt the greatest service the Fathers did in this undertaking, not because of its great importance for royal prestige, but because of the great peril and risk in which they placed themselves, for, if the mandarins had rejected this banner, all their rage and fury would have fallen on the Fathers. I leave aside the many letters and admonitions they sent from the court to the Lord Ambassador in which they instructed him how he should deal with the mandarins. On the same day on which the Lord Ambassador arrived at Cinquevan,[57] 4 leagues before the court, the Fathers came to visit him on the boats and told him the questions he would be asked at court. Finally, the fact that he entered the court with the honor and esteem to which I have referred was entirely due to the zeal and diligence with which the Fathers informed the Emperor, not only orally[58] but by a single printed book on the grandeur and power of the King Our Lord, of his realms and conquests, of the order of his tribunals, and of the nobility and ancient lineage of the house and family of the Lord Ambassador, so that he is better known in China today, and will be in the future, than among many people in Portugal. The Fathers attended him in his illness, coming every day from so far away to care for him and console him, staying with him from morning to night to give him pleasure. I accompanied him, and attended him for two years, saying Mass to which he was very devoted, for him every day, and, in the last sickness from which he died, I administered to him the Sacraments of which he was capable, not leaving him day or night, until he died in my hands. I closed his eyes, and composed his face, and, for the three days the seizure lasted (37), I waited with the Mass prepared to be able to give the Holy Viaticum, if the sickness should give him some interval in which he would come to his senses. This being the case, as all saw who were on the journey, it still is not sufficient to keep Bento Pereyra from saying that the Fathers ruined the embassy.

He [Bento Pereira de Faria] placed in the hand of the Lord Ambassador, before he entered the Palace, a memorial to be given to the Emperor in which the people of Macao cited the services they had done for the Emperor, expelling the Dutch and killing many of them, how many ships destroyed, how many pirates killed or turned over to the mandarins of the province, and other things of this kind. The Lord Ambassador asked the Fathers to give him their opinion, if it would be appropriate to present this memorial. The Fathers replied that, if this memorial was presented, it would do great damage to the embassy, and much more to Macao, and listed the following reasons. First, because, when Your Lordship in public tribunal by order of the Emperor was asked by the Tsung-tu of Canton if you came to discuss any business concerning Macao, you responded no, but you only came to present congratulations to the Emperor. The embassy was received solely on these terms,[59] and, if now you present this memorial, you deny what you said with notable discredit to your person. Second, because we have learned from the confidential Ko-lao [Grand Secretary], and from the Fourth Petty King, who are those who now govern the empire,[60] and our greatest friends and protectors, that the sea will not be opened until the affairs of I-kuan or the long-haired Chinese come to some conclusion, and, if Your Lordship submits this memorial, it is certain that the Emperor's decision will come out against them, and our mouths will be stopped so that we can never raise this matter again at any time. Third, because the greatest esteem in which this embassy is held both by the Emperor and by all (38) the empire is because it is known that Your Lordship does not come to ask for anything, nor to engage in trade, but only to offer congratulations to the Emperor; they praise this, and are astonished to learn that Your Lordship has come from so far away, and, if today you submit this memorial, they will know clearly that the giving of congratulations was a false front, and the memorial represents the sole purpose and real intent behind your coming, and for this reason all their esteem will be turned into contempt and ridicule, if they think that they have been fooled.

Fourth, this memorial contains deeds of war, and proof that the Portuguese are good soldiers and men of arms, and this is simply putting our heads on the executioner's block before the Tartars. This can be proved by what happened to the Dutch, which none of us knew, and it may be that the Dutch themselves do not know. The Tartars had conceded them two factories, one in Nanking [Chiang-nan], the other in Fukien, but on condition that they help them expel the long-haired Chinese who still held some islands which are near the province of Chincheo.[61] They accepted the arrangement, and came with 14 ships; supposing that their valor and military science were well enough known and respected, nevertheless on this occasion they were much helped by having the friendship of the Tartars, and showed their mettle and bravery with greater immoderation than ever, and, although they lost three ships, they defeated the enemy and occupied the islands, which they soon turned over to the Tartars. Who would have thought that this action would have led them to lose China forever and gain hatred and general abhorrence where they deserve respect and esteem? The Fathers said that the Tartars who returned from that war to the court entered putting their fingers in their mouths, a sign of great astonishment, crying in the streets, "Take care, take care—out, out with people who fight this way; this admitting them is putting fire in the bosom, taking tigers into the house." Thus, the Dutch, while they wanted to make themselves desirable friends, made themselves formidable and odious, and soon were treated so badly that not only were they not given the promised factories but they were expelled with a thousand abuses and affronts. Now let those of Macao see if it is good that they should cite the service of their arms and valor, when here at the court the Tartars live in fear of artillery that is in Macao, and have spoken of it many times!

(39) For these reasons, the Lord Ambassador decided not to present the memorial. At once Bento Pereira complained that the Fathers had frustrated them [and the Fathers responded] by saying,[62] "My Lord, we govern nothing here; the Lord Ambassador asked us our opinion and we told him what we understood before God; if the contrary seems better to you, do as you wish, and who will stop you? Do what you can, negotiate where you can, with the Emperor and his tribunals." Bento Pereira told of this opinion of the Fathers in Macao to make us hated, but did not cite the reasons on which the Fathers founded it, nor how many times speaking with the Emperor in private they had informed him of the tribulations of this City, not only by word of mouth, but also by a printed book which the Fathers gave me[63] and which I have here. It can well be seen that the Fathers spoke to the Emperor about Macao, for, when the Lord Ambassador spoke [of it], the Emperor replied that he already knew everything, and there was no one there who would speak of it besides the Fathers. Bento Pereira knows this well, but dissimulated because it does not serve his intention of making us hated. Nor does he speak of the favorable decisions as a result of which Macao did not have to move to Canton as they were ordered to do twice in Holy Week of 1666, and another so that they might stay within [the city] and the Circle Gate be opened. Who defended them against the two infamous memorials against this city by the Petty King, Tsung-tu, and Fu-yuan of Canton? How many friends do the Fathers have at court and in China, who are among the most powerful people, and favor Macao out of respect for them? If one has any doubt about this, he should ask Bento Pereyra and the people of Macao if the Fathers did not obtain these decisions for them, and whom the people of Macao have at court or anywhere in China who could do this favor for them. If they are favored by the confidential Ko-lao and the Fourth Petty King who now govern everything, by what did they gain it? What presents did it cost them? These two Princes, because they are the benefactors of Macao, expected some precious jewel from the Lord Ambassador, and when the Fathers saw that he did not have anything sufficient to send, they gave a precious clock to the Ko-lao in the name of the

Lord Ambassador, which we all saw when we went to visit him [the Ko-lao], and to the Petty King other curious items which the Fathers of Macao and Canton sent him, making up for this great lack with their poverty and diligence. When the Fathers told me of these presents which they sent in the name of the Lord Ambassador, I told them that, if they had gone to this expense in hope of some expression of gratitude from the people of Macao, they count it as lost; to which they replied, "The conservation of Macao means more to us than to anyone who lives there, and it means little if they thank us for all we do (40) to conserve them." Nonetheless it is astonishing, the ingratitude that covers their eyes so that they don't know their benefactors, to whom they have shown themselves so ungrateful that not only did they not write a letter to them on this occasion, but wrote many which I saw against them, in which they make them out to be their enemies. They also complain of the Lord Ambassador that they spent much on him, but he did not bring them the free trade for which they long; as if the decision were in the hands of the Fathers, or of the Lord Ambassador! They count the expenses of the embassy as lost, and do not realize that, even if the sea was not opened, still the respect and esteem with which the Emperor and the whole court treated the Lord Ambassador are roots that Macao has put down in China; for up to now the Emperor has not decided to put forth any edict against Macao, although many have solicited one, in order not to offend a King and a nation to whom so recently so many honors were done at his court.

My beloved Fathers will have observed that up to now I have said nothing of the fervor of the Christian communities in the areas I passed through, nor have I told some matters of edification which God Our Lord was pleased to bring about by means of this journey; but I left them for last on purpose in order not to mix the sacred and the profane. The first person I baptized on this journey was a little girl in Canton who was found abandoned in our street; she must have been four or five years old; she lived a few hours after the baptism, and then went to Heaven. Our soldiers, who are much given to this form of piety, dressed her in white silk *ling* [damask] with many flowers and roses of gold and silver paper, and went to bury her in a coffin which they bought for her, with a big escort. I baptized another boy child of about two years, son of a sailor on our boat, and he soon died, and the sailors gave him the same burial. Some of the Chinese were edified; others said that the Portuguese were crazy, for they spent money to no purpose.

From Canton to Peking in the going and returning, the number of people I baptized, children and adults, exceeded two hundred. But because God brought some to baptism by extraordinary means, I will mention two cases that gave me particular consolation. After we entered that famous river which, because of its greatness, the Chinese call "sons of the sea,"[64] and it is about a league wide, a severe windstorm came upon us, and with it another storm of cries from our kitchen, and the reason was that our cooks were frying fish, and the sailors demanded of them [that they stop, saying] that that part of the river was governed by a great serpent that lived[65] beneath its water, which, smelling fried fish, soon became irritated and moved the waters, making a great storm, and for this reason we were suffering. The one who played the largest part in this complaint was a very old sailor. (41) My slave was there, and just laughed and poked fun at the sailors, and especially at the old one, who said to him, "Here we are all so afraid, and you're laughing?" My slave took him by the hand and began to preach to him, and, although his talent for this calling was none too great, for he was of the Timor caste, still he knew the language very well, and spoke as follows: "You all are great fools who go around worshiping pagodas and serpents, which don't count for anything, and don't know that there is only one God who rules in Heaven and on earth, and all other things, even demons, deities, and serpents, are his servants and subjects and obey his commands, just as in China there is only one emperor.[66] He did not speak to him of any more profound concepts, but these were enough to be planted in that good old man, and left him so inclined to our Holy Faith he came to ask for his book of devotions, for he was literate in Chinese, and a few days thereafter he came to speak with me, saying that he wanted to become a Christian and to be baptized. I instructed him and kept him with me until we reached Nanking, where I commended him to the catechist of the church, and a few days thereafter I baptized him.

In the city of Kan-chou, God brought another to Faith in Him in this manner. This one was most devoted to the pagodas, and, as he told me himself, no one in that city had more beautiful and gilded ones, and his altar very clean, and always with incense by day and night. It happened that there passed through that city a *tsung-ping* [brigade general] which is like a colonel of infantry,[67] and he lodged his soldiers in it. By chance, a Christian soldier was lodged in this man's house, and, after he had given him supper, took him to a lodging where he had made a bed for him. The soldier took from his pack an image, but he couldn't tell me of what saint, and hung it over the head of his bed, and they both went to bed. In the night the owner of the house dreamed that his pagodas one after another with great sadness and grief, got down from the altar[68] and were leaving by the door. He soon came running to the door and called to them saying, "What is it that makes you leave me? I've served you with such care for so many years, and now you flee and abandon me?" The pagodas answered him briefly, "You have here a great official; we can't be in the same house with him." He got up early in the morning and went to say good morning to the soldier, and, seeing again the image on the head of the bed, asked him what it was. The soldier answered that he was a Christian who worshiped no pagoda but only one true God, and that the image he had there was of a very holy man, who was very pleasing to the true (42) God, and, because of this, the same God worked through him great marvels in the world. With this reply he came to understand the dream, and said, "Without a doubt this is the official of whom my pagodas were afraid last night." He told him of the dream, and declared that he wanted to become a Christian and also serve that God that had such great officials that his pagodas fled from them. He received Holy Baptism and remained so fervent that soon he brought others to be baptized.

I do not know what foundation there is for the rumor which some who are jealous of the Company [the Jesuits] have spread, saying rashly that the missionaries in China do not preach Christ Crucified; and because a particular case, full of Grace and relevant to this subject happened to my own slave, I will not omit to relate it. We arrived in Nanking on Palm Sunday and were there until the Wednesday after Easter, waiting for the completion of some painting on the boats in which we would go. On Good Friday, some Christians came to invite my slave to go worship the Holy Christ, which was on a rich cushion, in the form that is customary in a church on that day, and the church was furnished with many cloths of damask and other silks, thanks to the donations and the great piety of Lady Agatha, wife of the former Viceroy of this province.[69] The slave came to the church where he found the Christians gathered on their knees in prayer in great silence, and saw that one by one they went to the foot of the Holy Crucifix, and kneeling let their robes fall from their shoulders and inflicted on their backs a heavy discipline, and then with many tears kissed the feet of the Holy Image. The slave was much struck by this novelty, for he was not accustomed to inflict discipline on his back and to do it so extensively. But also he did not dare to omit doing what the others were doing, for, because he was a slave of the Fathers, they respected him as a master, or at least would pay attention to what he would say about that ceremony.

Finally his turn came, and he inflicted discipline like the others. Later he came to the boat and described what had happened with much grace, for he was a simple soul, and told of it among the hardships and adversities of the voyage. But I said to my companions and also to the Lord Ambassador, "See, gentlemen, do the Chinese know Christ Crucified? And do the Fathers give them knowledge of Him? And tell me in what part of the world He is worshiped in this fashion?" The falsity of this calumny also is demonstrated by one of the chapters written against the Fathers by the impious blasphemer and cruel persecutor of the Law of God, Yang Kuang-hsien, in which he said that the Fathers proclaim as God a man who died crucified for wanting to rebel against foreign rulers,[70] and, in order to make this sin more hideous and abominable to the temperament of the Chinese, in the same book in which he wrote these chapters he had painted the Christ Crucified, and this lost soul also proclaimed it in the tribunals when he persecuted [Christ's] ministers, as also (43) happened to the Apostle in Rome: Quidam quidem et propter invidiam, et contencionem, quidam autem, et propter bonam voluntatem Christum praedicant. But the Fathers console themselves, and rejoice greatly along with the same Apostle to see Christ known and proclaimed in whatever way it may be: Dum omni modo sive por occasionem, sive per viritatem Christus anuncietur, et in hoc gaudeo, sed et gaudebo, Philippians 1:15.[71] But the world already is freed from this monster, for not long before we arrived at the Court he died, spilling his entrails out his back by horrible abcesses that afflicted him, a death detested among the Chinese.[72] In Nanking I did not go to say Mass in the church, but on the boats, and on them I gave confession and communion to the Christians, for I was told that I should not go to the church in those days, for that city was Yang Kuang-hsien's home and he had many relatives there, and there were many Muslims, cruel enemies of the Law of God.

People will want to know how I could give confession, not knowing anything of the language. The system was as follows. Seeing myself pursued all along the route by Christians who came from very far to seek me out so that they could confess themselves, and that they were most disconsolate when they found that I did not know the language, I asked a Chinese learned man whom the Fathers had placed at the service of the Lord Ambassador to prepare papers when necessary to make for me a confessional register in Chinese letters, placing at the top those there could be under each precept, and in a separate place the number of times up to five, and above that it was to be indicated by the fingers. Beside the Chinese letters I put the Portuguese. They confessed the kind of sin, and then quickly looked for the letters that stated the number of times, and in this way they made it clear. Many were not content, but called my slave, and through him stated all they had on their consciences, even though I told them they were not obliged to do so. These confessions were of great importance for some, for when I returned from the court I found them already dead; among these was the master of the church of Kan-chou, who, after confessing himself and taking communion, accompanied me to the boat, and from the boat went home with a fever and illness of which he died.

So that the women also could take advantage of this opportunity, they gathered in a house and quickly sent for me, and there they said their confessions and then I said Mass for them and gave them Communion, but never did a single woman confess on paper, although many of them were literate, but all confessed through interpreters despite the declaration I made to them that they did not have that obligation except on the point of death; to which they replied that they did not want to miss this opportunity because they did not know if they would have another in their lives. I was greatly edified by this opinion and by the good reputation of the Fathers in China, where the women live in their houses in such (44) seclusion that they may not even speak with their brothers, but in me, who went in lay dress like any one of the gentlemen in waiting of the embassy, they placed so much trust, just because they knew that I was of the Company, even though the Fathers had not given me letters to the Christian congregations—because I would be en route and did not know the language they thought they would be of no use to me—still the Christian Chinese whom the Lord Ambassador had in his service quickly informed the people in the cities through which we passed that there was a Father of the Company. Others came to search for me only because our boats were flying flags with the Cross of the Order of Christ.[73]

In Nanking I was summoned to the house of T'ung Lao-yeh, [His Excellency T'ung], a person of the highest standing, a very close relative of the Emperor, who had passed through some of the highest positions in China, from Fu-yuan in Kanchou, which is governor of the city, to Tsung-tu in Fukien, which is viceroy, and Tsung-tu again in Nanking, and now is destroyed, all his positions lost, after having spent almost all his silver, for in this persecution they accused him of being a friend and supporter[74] of the Fathers, and of having built three large churches to the true God. This great nobleman has much esteem for the Holy Law, and a better-than-ordinary knowledge of spiritual matters, and does not want to have in his house any servant who was not a Christian, but, despite all this, and being already of the dignity, I should say the age,[75] of seventy-three, he had not yet come to receive Holy Baptism, because he cannot separate himself from women; he leaves this matter as he says for its time; it seems that he trusts in his constitution, which is strong and robust. He received me in his house with the same honor and polite usages as to his equal, another viceroy, and, after giving me tea and sweet pastries in the company of his sons, he entertained me in the garden until night came, and then quickly took me to the house where was Lady Agatha, a most fervent Christian, his principal wife, with many others of the same family and from outside it, for this nobleman is the support of that Christian community, and the three churches he built are the only ones that, out of respect for him, have been preserved as churches until today, and the rest the officials took for themselves and profaned. But now they in turn have restored them, for the Fathers at the court have obtained dispatches from the Emperor that the Fathers exiled in Canton should return again to their churches; the edict already has come from the court, and the order has been given in Canton.[76] I spent most of the night hearing the confessions of these ladies, which all were through interpreters, and before dawn I said Mass and gave them Communion; after the Mass I baptized two girls of the same household. In the morning, it being already broad daylight, I said another Mass and gave Communion to the men, for to have both men and women in the same house or church is a thing not tolerated in China. These spiritual exercises being done, the Lord T'ung Lao-yeh invited me to dine, and, after showing me many favors there, he gave me permission to return to the boat, which already was setting sail, and (45) soon thereafter sent a splendid gift of things to eat for all the people of the embassy. On all this route, wherever there were Christians, the boat went its way on the river, and I went by land from place to place administering the Sacrament to the Christians; where there were none, I went back to the boat. These are, in summary, the events of this journey told in all the truth that is possible in human affairs, for I took part in everything, and was an eyewitness. For this reason I deserve some credence, and also pardon for the crude style, and the bad temper of which I had a little when I wrote it.

Father Francisco Pimentel

appendix b

Ferdinand Verbiest, S.J., on the Embassy of Bento Pereira de Faria, 1678

Note: This is an extract from an annual letter for 1678–1679 of the Chinese Vice-Province of the Society of Jesus; it is preserved in two copies in the Archivum Romanum Societatis Jesu, ARSJ/JS 117:161–182 and 183–202. According to a personal communication from Father Francis A. Rouleau, S.J., 21 January 1969, the former is in the hand of Verbiest. Neither copy bears a signature or date; internal evidence suggests that it was written at the end of 1679. The passage transcribed is on fol. 164–164v and 185–185v. A summary of this passage in Thomas Ignatius Dunin Szpot, S.J., "Collectanea Historiae Sinensis, ab anno 1641 ad annum 1700," ARSJ/JS 104:306–307v, has been consulted to clarify some points. This translation is based on one prepared for me by Celeste Anderson, later checked and corrected by Lee Reams and by John W. Witek, S.J. Verbiest acknowledged that his Latin was rusty after decades in China.[1] At a number of points the Latin text probably could not be understood without a good knowledge of court and embassy routine.

Thereafter, upon the occasion of the arrival from Macao of the Ambassador D. Bento Pereira with a lion as a royal gift, the Emperor presented the companions in Peking with various honors and favors, which I only mention briefly here, for Father Philip Grimaldi wrote an extensive report on all the events of the embassy and sent it to Europe; I refer the reader to it.[2]

So much was the Emperor pleased with the present of the lion, that by special favor and grace the Princes once were granted permission to see it.[3] Moreover, he wished that our Peking Fathers be included in the small number whose standing made them worthy of being present on this occasion. Another instance of rare favor was that, he wanted to be present himself with those who were viewing the lion, and wanted [the Fathers] to stand close to him and to speak with him familiarly.

When the Emperor admitted the Lord Ambassador into his presence for the first time, he ordered a royal banquet to be prepared for him in one of his greatest halls, where he seated himself on his royal throne and, dining, wished to honor the Lord Ambassador with some dishes from his royal Table. At the same time he also presented a similar honor to our Father Ferdinand, who was ordered to accompany the embassy there, stood by during the ceremonies, and served as interpreter both in Chinese and in Manchu.

After this banquet, wishing to honor again the Ambassador and his son, Jeronimo Pereira, he ordered them to be invited separately to Chin-shan, that is, Gold Mountain, a place of wonderful pleasantness, set aside primarily for the recreation of the Emperor, so that they might see it. And here, another royal banquet being prepared, he ordered all four of our Fathers to be invited at the same time, wishing indeed that they share in this royal favor.

In the Board of Ceremonies, which has jurisdiction over such embassies, they did not wish to use the interpreter who had come with the Ambassador to translate into Chinese and Manchu the letter from the Most Serene Prince of Portugal, but they used our Fathers Ludovico Buglio and Ferdinand Verbiest. After the letter had been translated and Father Ferdinand had been summoned by him, the Emperor personally ordered him to open the most elegant silver box in which the letter of the Most Serene Prince of Portugal was offered and to read it aloud in the Portuguese in which it was written, publicly displaying through this the confidence he had in our Fathers.

He also ordered the Board of Ceremonies that our Fathers were to be allowed to go to visit the Ambassador whenever they wished.

However, this embassy was undertaken by the Macaenses chiefly in order to obtain free commerce with the interior of China and permission to sail in any direction they wished.[4] This goal happily was achieved this year[1679]; in this Father Ferdinand was the chief agent, working through the Ko-laos, supreme royal ministers, and through the Emperor by way of the young men whom he sometimes sent to our house. The same City of Macao in several letters of thanks to the Fathers has testified openly that it is indebted to Father Ferdinand for all the favors it has received from the Emperor; these letters the said Father still has in his possession. And the second Ko-lao, named Ming-chu,[5] to whom the Father went to give thanks for the favors obtained, openly said these words to him: "It is entirely to your credit that the City did not have to change its location, and its inhabitants never were ordered to do so (this is from the time of the persecution by Yang Kuang-hsien, and after), and also it is to your credit that it now has obtained this commerce and free navigation. You have opened the way here not only for the propagation of your astronomy, but also your religion." He said these things. These I wish to bring forward here, so that posterity may know why and how much our Society did for the good of this City of Macao.

For the rest, not long afterward the lion died, and the Emperor sent the same man who had counseled our Fathers to ask whether it seemed good to them to bury it without those [parts] which it might be permitted to remove from its body on the pretext that they have medicinal virtues, as it is commonly said to have in various parts of its body. Thus he wished to set him apart from all others in his esteem by this distinction, that he wished that nothing be done until he received the Fathers' response.

Occasione deinde advenientis Macao Prolegati D. Benedicti Pereira cum leone munere Regio, variis honoribus ac favoribus Imperator affecit socios Pekinenses, uti hîc breviter solum attingam. Nam extensam de toto Prolegationis huius successu relationem P. Phillippus Grimaldi scripsit, etiam in Europam praemisit, ad quam proinde lectorem remitto.

Adeo aestimavit et gavisus est oblato hoc leone Imperator, ut ex speciali favore et gratiâ vix ipsismet Regulis semel illius videndi copiam fecerit. Voluit tamen Patres nostros Pekinenses intrare in hunc numerum paucorum, quos hâc gratiá suâ dignari statuerat; additâ insuper hâc alia rari favoris circumstanciâ quod ipsis leonem spectantibus ipsemet praesens adesse, familiariter colloqui, sibique proximos adstare voluerit.

Quando Imperator D. Prolegatum primâ vice in praesentiam suam admisit, epulum eidem Regium instrui iussit in unâ é maioribus aulis suis, ubi ipsemet in Throno suo regio considens, et prandens, aloquot discis è mensa suâ regiâ honorare voluit D. Prolegatum. Quem eundem tamen honorem eodem simul tempore praestitit Patri nostro Ferdinando, qui eiusdem jussu huic excipienda legationis solemnitati assistebat, et interpretis tum Sinici tum Tartarici vice fungebatur.

Post regium hoc epulum honore iterum volens afficere Prolegatum hunc cum filio suo Hyeronimo Pereira; invitari illos seorsim jussit ad Kīn Xān, id est, aureum montem, locum mirae amaenitatis, recreando in primis Imperatori destinatum, ut eum viderent. Atque hîc, dum alterum iterum epulum regium instruitur, advocari simul jussit omnes quatuor Patres nostros, volens scilicet etiam hos regii huius favoris sui fieri participes.

In Tribunali Rituum supremo, ad quod eiusmodi legationes spectant, noluêre uti interprete, qui cum Prolegato advenerat, ad vertendas idiomate Sinico, et Tartaro litteras Serenissimi Principis Lusitaniae; sed usi sunt Patribus nostris Ludovico Buglio, et Ferdinando Verbiest. Versis litteris, et accersito ad se Patre Ferdinando, Imperator eundem jussit coram aperire capsam elegantissimam eamque argenteam, in qua serenissimi Lusitaniae Principis litterae offerebantur, easque alta voce legere ipsomet idiomate Lusitano, quo scriptae venerant, palam per hoc faciens magnam, quam erga Patres nostros habet confidentiam.

Mandavit item Supremo Rituum Tribunali, ut quacumque demum hora nostri patres adire ac invisere vellent D. Prolegatum; liberum eisdem accessum permitterent.

Fuit autem haec legatio a Macaensibus eo praecipuè fine procurata, ut liberum cum interiore Sinâ commercium, et quaqua versum navigandi facultatem obtinerent, uti sub anni huius finem feliciter etiam obtinuêre, agente cum primus P. Ferdinando tum apud Colaos supremos regni ministros, tum apud Imperatorem per ephebos suos, quos domum nostram identidem mittit. Quod Imperatoris beneficium Patri Ferdinando praecipue se debere etiam ipsa civitas Macaensis datis iam aliquoties ad Patrem Litteris gratiarum actoriis, quas idem pater penes se adhuc habet, aperte testata est. Et secundus Colaus, cognomento Mîm chū, cui Pater gratias pro obtento hoc beneficio acturus ipsum adiisset, aperte eidem haec verba dixit: Quod civitas Macaensis olim non fuerit jussa mutare sedes suas, et quot Macaenses cives non sint olim amandati, (id est tempore persecutionis yâm quām, siēn, et post) hoc totum tuum est meritum, nunc autem quod obtinuerit hoc commercium, et liberam navigationem, hoc etiam tuum est meritum: Tu enim non tantum Astronomiae, sed etiam Religioni tuae hîc propagandae viam aperuisti: haec ille. Quae hic adferre volui, ut posteritas sciat quid et quantum nostra societas ad bonum civitatis huius Macaensis in hoc operata sit.

Caeterum mortuo haud multo post leone, misit idem Imperator, qui Patres nostros consulerit, num scilicet ipsis videretur sepeliri leonem absque eo quod ulli permitterentur detrahere vel tollere quidquam de eiusdem cadavere praetextu seu titulo virtutis alicuius medicinalis, quam habere communiter dicitur in variis sui corporis partibus; volens scilicet differenti illum ab omnibus alijs aestimatione secernere: neque executioni dari quidquam voluit prius quam Patrum responsum accepisset.

appendix c

Three Poems on the Lion Brought by the Portuguese, 1678

By long stages, bearing gifts, they come from farthest west,
As tribute bringing a spirit-beast from east of the clouds of the sea.
In the Imperial Park it gazes like lightning over the crags.
At the Heavenly Gate it roars like a wind of ten thousand li.
In years past it defeated the enemy, tamed the elephant;
Today in the Park we saw it attack a bear.
The ministers rejoiced to see the chief of the furry tribe;
For the first time believing its bones matched legend's golden beast.

—Wang Hung-hsu[1]

A lion has come
From the Western Ocean,
It lips curl back, its claws are hooked, the head-hair yellow.
Where is this land of Libya?

You tread on gleams of light,
You leap over drifting clouds.
Splendid are the misty airs of the Imperial Park!
The bear is tamed, the tiger crouches, they are not your peers.

Ah, you lion!
Are you the one that eats a sheep a day?
The barbarian officials chatter ch'ih-ch'ih,
Glossy black without dyeing!
They came from oceans of poisonous fog,
Now to the south,
Now to the north,
Ten years on their way, bringing tribute;
How can we say they are late?

In the Bright Hall the offering is to the Earth Lord.
Singing the "Red Dragon,"
Drumming the "Vermilion Egret."
The strange-haired lion dances in time to the music.
Receiving rare gifts,
Treating men from afar as ministers,
Wherever sun, moon, and stars shine, none does not come to court.
The phoenix is on the mountain,
The unicorn is in the marsh.
The Prince rejoices;
May he live ten thousand years.

—Li Ch'eng-chung

The ancient emperors, careful of their virtue, fixed an Office of Interpreters.
The great lords were pacified, so too the wild tribes.
Their tribute gifts were set down in the royal records,
Their precious globules listed in the Hall of Guests.
River gold and western plumage were offered up on high,
The barbarian music was heard in the Bright Hall.
The three divinities responded, the hundred spirits joined,
All kinds of beasts wandered and soared together.
In the wu-wu year, seventeenth of K'ang-hsi,
The fame of divine arms had spread far beyond the seas.
At the five high altars the teachings are followed, the white spirit-bear comes;
At a thousand gates the splendor of the yellow dragon is seen.
Black birds fly hsi-hsi, shot but flying on;
The fowl rise up and roost in the forest branches.
From many regions come bearers of presents, lined up in majestic rows;
All the way from the north and east seas, and the flowing sands of the west.
It's not that rare animals are prized, but the men from afar are cherished.
The tiger cages are checked and moved to the Imperial Park;

The falcon cages, when autumn comes, are moved to the Southern Park.
The court ministers follow behind, wanting to write poems.
By Imperial Grace, they're given time to come see.
Round eyes, proud nose, muscles of awesome power
Fierce and quick to move, whether male or female.
We only regret its mane is not all curly;
Its long tail swishes like loose silk.

In color as if dressed for Audience,
Its fearsome fangs it need not use.
Then all the beasts were in their cages,
Wooden bars, railings of spears, along the Front Garden.
The bears moved away from the road, and dare not hold their ground;
The tigers and leopards claw at their cages, roaring now and then.
The green *luan* bird and the red *chiao* face each other by the railings;
Wild boars, wild horses, gambol in the fields.
Hero generals of the Han would not have been frightened;
With Chu Hai the tiger-tamer at your side, who would fear?

—Mao Ch'i-ling
(selected passages)

appendix d

Gifts and Food Allotments from the Ch'ing Court

The first of the following charts shows the careful categorization and standardization of imperial gifts to monarchs sending embassies and to ambassadors and their suites, including the increased gifts to the Pereira de Faria embassy which were one important sign of imperial favor. The presents to the "Holland King" of course went to the Governor-General at Batavia. The figures for silks are summaries of more detailed lists in the Chinese sources, in which a few further status distinctions seem clear and more might be seen by an expert on sumptuary rules in the use of silks. For example, only the King and the Ambassador received a piece or two of satin decorated with four-clawed dragons (*mang-tuan*). In his report, Vincent Paats described the silks received as "old-fashioned," as types current in trade years before.[1] A brief survey of lists of presents to other kings and ambassadors reveals few obvious universal patterns. Probably the gifts of satins for the king were the most stereotyped and taken the most seriously by the Ch'ing authorities. The De Goyer–De Keyser embassy had received only 30; in 1684 the 50 given to Pereira de Faria became the norm for most embassies, but not, apparently, for Paats.[2] For the Paats embassy, the gifts to "King" and Ambassador were the same as for Van Hoorn. The Chinese texts show that presents for anomalous individuals were decided on in terms of classifications of personnel. Thus, Joan van Hoorn and Pereira de Faria's son each received gifts equal to those to a soldier, and Louis de Keyser as Assistant Ambassador equal to those for a senior official.

The second chart shows how food rations were dispensed in Peking with obsessive bureaucratic precision, down to tenths and twentieths of *liang* of pepper. It is highly probably that the food rations for Pereira de Faria were the same as for Saldanha and that those for Paats matched those for Van Hoorn; they were recorded as precedents to be applied to later embassies from the same country. It is interesting that the soldiers received pork, the most common Chinese meat, while the Ambassador received a more varied meat diet, more in keeping with the tastes of the Manchu rulers. I suspect the assistants received more meat than the chart shows, and that an item or two were lost in the copying of the original records into our Chinese sources.

The charts are based primarily on the K'ang-hsi *Collected Statutes; the Rules and Precedents of the Board of Ceremonies* are the sole source for some of the lists of gifts, and it and other sources repeat much of this information with small editorial changes and some copying errors.[3] All this information except the presents to the Portuguese embassies is corroborated by European sources.[4] The numbers in parentheses for the Saldanha embassy represent the totals of the statutory presents recorded in the Ch'ing compendia and the informal gifts bestowed by the Emperor on August 31; the latter are known only from European sources. There are a number of difficulties as to the exact identity of items listed and units of measure, but the correspondence is surprisingly close, and it is clear that the gifts and food rations described in the Chinese offical sources actually were given. The Dutch list of food rations lists as single totals the rations for the Ambassador, Nobel, and Joan van Hoorn. The *Collected Statutes* lists figures for "the Ambassador and Assistant Ambassador together", but, for all except the sheep and the fruit, the quantities in the Dutch list are much larger, and usually exactly three times those in the Chinese lists, suggesting that, in fact, the Ch'ing officials gave rations equivalent to the Ambassador's to Nobel and to Joan van Hoorn.

The lists also include fragmentary references to presents and rations for the Ch'ing officials and interpreters who accompanied the embassy. This reminds us that the whole focus of these records is on precedents needed by Ch'ing bureaucrats. These presents and rations were not illusory, but the detailed records of them contributed to later Ch'ing illusions of the reality of a system of unilateral bureaucratic control.

Imperial Gifts

Recipient	King			Ambassador			Senior Officers, each			Junior Officers, each	Soldiers, each		
Embassy	VH	S	PdeF	VH	S	PdeF	VH	S	PdeF	VH	VH	S	PdeF
Satins	40	40	50	14	15(25)	15	7	6(12)	7	4	0	0(4)	0
Other Silks	40	40	50	12	10	15	6	6	8	4	4	6	8
Silver, taels	300	300	300	100	100	100	50	50	50	40	15	20	20

Daily Food Rations in Peking

	Ambassador		Assistants, Interpreters, each		Soldiers, each	
	VH	S	VH	S	VH	S
Milk, pots	1	1	–	–	–	–
Pork, *chin*	–	–	–	–	1.5	1.5
Sheep	0.5	1	–	–	–	–
Goose	1	1	–	0.5	–	–
Chicken	1	1	–	–	–	–
Fish	1	1	–	–	–	–
Fruit, pieces	35	44	–	–	–	–
Grapes, *chin*	1	1.25	–	–	–	–
Dates, *chin*	1	1.25	–	–	–	–
Tea, *liang*	1	1	0.5	0.5	–	–
Flour, *chin*	2	2	1	1	0.5	0.5
Beancurd, *chin*	2	2	1	1	–	–
Pepper, *liang*	0.1	0.1	0.05	0.05	–	–
"Clear sauce," *liang*	0.6	0.6	0.4	0.4	–	–
Soy sauce, *liang*	0.6	0.6	0.4	0.4	–	–
Sesame oil, *liang*	0.6	0.6	0.4	0.4	–	–
Vegetables, *chin*	3	3	1	1	–	0.125
Wine, pots	10	6	1	1	5/15	6/19*

All also received rice and salt; supplies of lamp oil also are recorded for the Saldanha embassy.

* 5 pots for 15 soldiers; 6 for 19.

appendix e

Gifts Brought by the Embassies

This chart summarizes the information available to a reader of Chinese bureaucratic compilations, with clarifications from full European lists of the Saldanha and Paats presents and fragments of Dutch information on Van Hoorn's.[1] Not included in the chart is the lion brought by Pereira de Faria, which probably made a greater impression in Peking than all the presents brought by any one of the other embassies. Anomalies such as Van Hoorn's presents in his own name and Saldanha's presents to the Emperor are recorded without comment. For more regular embassies such as the Korean and Ryukyuan, these records were treated as precedent-setting, but the only case of that here is the order after the Paats embassy that in the future the Dutch should bring only thirteen kinds of goods: horses, coral, mirrors, amber, cloves, sandalwood, camphor, guns and flints for them, woolens, satins, gold-woven rugs, and clocks.[2] The urge to bureaucratic routinization and the impulse to reduce presents in the name of "being kind to those from afar" here were held in a nice tension with fascination with their exotic animals and manufactures. Horses and guns could be justified in terms of use in war. The admissability of clocks echoes a fascination first exploited by Ricci, later by generations of Canton traders selling their "sing-songs."

	Swords, daggers	Guns	Woolens, satins (pieces)	Cottons (pieces)	Coral (strings)	Amber (strings)	Cloves (*chin*)	Pepper (*chin*)	Sandalwood (*chin*)	Elephant Tusks	Incense, rose water, etc. (bottles)	Miscellaneous
Van Hoorn, to Emperor	8	17	73	296	2	4 pieces	500	500	2000	5	–	4 horses, 8 sets saddles and harness, 2 small oxen, cart, 2 big mirrors, 1 glass dish, 15 cushions, globe, celestial sphere, 4 carpets, glass lantern
Van Hoorn, to Emperor, in Ambassador's Name	2	6	10	–	4	1 piece	–	–	–	–	20	3 pieces fine metal work, 4 large bird eggs (ostrich?), suit of armor, 7 copper models, 4 mirrors, 2 walrus tusks
Saldanha, to Emperor	1	–	3	4	1	6	1 box	–	–	10	11	Portrait of King of Portugal, box made of gold and amber, calambac wood, 4 rhinoceros horns, 1 carpet
Saldanha, to Empress	–	–	–	4	1	10	1 box	–	–	–	1	1 mirror, 1 carpet

	Swords, daggers	Guns	Woolens, satins (pieces)	Cottons (pieces)	Coral (strings)	Amber (strings)	Cloves (*chin*)	Pepper (*chin*)	Sandalwood (*chin*)	Elephant Tusks	Incense, rose water, etc. (bottles)	Miscellaneous
Paats, to Emperor	46	80	45	364	1	24 pieces	30	–	–	5	1 box	2 mirrors, 4 carpets, 1 clock, 1 copper lantern, 1 candelabra, 580 glasses, camphor, nutmeg, 3 telescopes, 3 ship models, 2 bottles wine
Paats, to Emperor, in Ambassador's Name	6	13	–	80	–	–	–	–	–	–	–	1 silver plate, 1 silver vase

Abbreviations

The following short forms and abbreviations are used to refer to manuscript sources and a few repeatedly used book titles. Colons are used to separate the designation of the manuscript volume from that of the page in it, or of the volume or chüan of a printed work from the page.

AHU — Arquivo Historico Ultramarino, Lisbon

Ajuda/JA — Biblioteca da Ajuda, Lisbon, collection entitled "Jesuitas na Ásia"

ARSJ/JS — Archivum Romanum Societatis Jesu, Rome: Archives of the Japan-China Province

Goa — Historical Archive, Panjim, Goa. "Monções" indicates number in the series of "Livros das Monções", "Monsoon books", the basic collections of correspondence between Goa and the other parts of the Indies on the one hand and Lisbon on the other. Other volumes are cited by the Archive's index number.

HTSL — *Ta-Ch'ing hui-tien shih-li*

LPTL — *Li-pu tse-li*

Palha — Houghton Library, Harvard University. Palha collection of manuscripts. I refer to a set of documents on Portuguese embassies to China bound in the back of D. Luis da Cunha, "Memorias da Pas de Utrecht".

SL:KH — *Ta-Ch'ing li-ch'ao shih-lu,* K'ang-hsi reign

SL:SC — ibid., Shun-chih reign

TW: — *T'ai-wan wen-hsien ts'ung-k'an,* Taipei, 1958 et seq.

VOC — Archives of Verenigde Oostindische Compagnie, Algemeen Rijksarchief, The Hague. For explanation of various series see my *Pepper,* pp. xii–xv. The "VOC" numbers used here replace the "KA" numbers used previously. Almost all correspondence in the Dutch East India Company was between governors or chief merchants and their councils. I list these by location; thus, for example, "Foochow to Batavia."

Notes

ONE: CONTINUITIES AND ROUTINES

1. This description is based on Olfert Dapper, *Gedenkwaerdig Bedryf* . . . (Amsterdam, 1671), pp. 354-356. See Chapter 2 for a full account of the Van Hoorn embassy and a discussion of Dapper's work and other sources. The requirement to dismount and the worry about hats falling off are noted in the records of the Saldanha embassy; see Chapter 3. For the ceremonies of the ordinary audience, see *Ta-Ch'ing hui-tien,* K'ang-hsi, Chapter 41; ibid., Kuang-hsu, 27:1-3; Christian Jochim, "The Imperial Audience Ceremonies of the Ch'ing Dynasty," *Bulletin of the Society for the Study of Chinese Religions,* 7:88-103 (Fall 1979). One's understanding of these ceremonies is immensely enhanced by a chance to see the places where they occurred; I am grateful to the Committee on Scholarly Communication with the People's Republic of China and the Academy of Social Sciences of the People's Republic of China for giving me such an opportunity in June 1979. Information on palace dimensions and other details is from charts in Osvald Sirén, *The Imperial Palaces of Peking* (Paris and Brussels, 1926), Vol. 1, and Chinese Academy of Architecture, *Ancient Chinese Architecture* (Peking and Hong Kong, 1982), pp. 138-142.

2. John E. Wills, Jr., *Pepper, Guns, and Parleys: The Dutch East India Company and China,* 1662-1681, (Cambridge, Mass., 1974), pp. 204-206.

3. Thus, for example, there have been modern studies of three of the four embassies in this book—see Note 6 below—but no coherent studies of the rest of Sino-Dutch and Sino-Portuguese relations in this period until my *Pepper* and the sections on Macao in the present book. Other embassies that have received detailed treatment in Western languages are those of Tomé Pires (1517-1524), Macartney (1793), and Titsingh and Van Braam (1795). The classic study on the whole system is J. K. Fairbank and S. Y. Teng, "On the Ch'ing Tribute System," *Harvard Journal of Asiatic Studies* 6:135-246 (1941), reprinted in Fairbank and Teng, *Ch'ing Administration: Three Studies* (Cambridge, Mass., 1960). The most sophisticated set of case studies and theoretical statements is John K. Fairbank, ed., *The Chinese World Order* (Cambridge, Mass., 1968).

4. In addition to the case studies in Fairbank, *Chinese World Order,* see, for example, Sarasin Viraphol, *Tribute and Profit: Sino-Siamese Trade, 1652-1853* (Cambridge, Mass., 1977), and Henry Serruys, C.I.C.M., *Sino-*

Mongol Relations During the Ming, II, The Tribute System and Diplomatic Missions, Mélanges Chinois et Bouddhiques, 14, Brussels, 1967.

5. The only other substantial record I know of of an embassy from outside the East Asian culture area is Hafiz-i Abru's famous account of the Timurid embassy of 1420. K. M. Maitra, tr. and ed., *A Persian Embassy to China; Being an Extract from "Zubdatu't-tawarikh" of Hafiz Abru* (Lahore, 1934; reprint, New York, 1970).

6. The most important studies are those by Lo-shu Fu and Luciano Petech, cited in Chapter 3, and Jan Vixseboxse, cited in Chapter 5.

7. For general descriptions of the archives consulted, see John E. Wills, Jr., "Early Sino-European Relations: Problems, Opportunities, and Archives," *Ch'ing-shih wen-t'i,* 3.2:50-76 (December 1974).

8. *[Ta]-Ming hui-tien* (1587; reprint, Kuo-hsueh chi-pen ts'ung-shu, No. 80, Taipei, 1968), ch. 58, 109-113; *Ta-Ming chi-li* (1530), chüan 31. At present, there is considerable interest in Chinese state ceremony among scholars writing in English, and much more sophisticated interpretations should be possible in a few years. My own approach is outlined in John E. Wills, Jr., "State Ceremony in Late Imperial China: Notes for a Framework for Discussion," *Bulletin of the Society for the Study of Chinese Religion,* 7:46-57 (Fall 1979).

9. *Ta-Ming chi-li,* 31:1, 31:9b; James Legge, tr., *The Chinese Classics,* 5 vols. (reprint, Hong Kong, Hong Kong University Press, 1960), 3:345-350; *Ming-Ch'ing shih-liao* (Peiping, Shanghai, and Taipei, 1930 et seq.), 3rd collection, 336-338b, at 338; *Ming-shih,* 332.5b, cited by Joseph F. Fletcher, "China and Central Asia, 1368-1884," in Fairbank, *Chinese World Order,* pp. 206-224, 337-368, Note 80. *The Ming-Ch'ing shih-liao* passage is translated in part in Lo-shu Fu, *A Documentary Chronicle of Sino-Western Relations (1644-1820),* Association for Asian Studies Monographs and Papers, 22 (Tucson, 1966), pp. 11-12. Professor Fu's pioneering work remains an invaluable reference tool for all students of early Sino-Western relations.

10. Legge, 3:92-151.

11. Herrlee G. Creel, *The Origins of Statecraft in China. Volume One: The Western Chou Empire* (Chicago and London, 1970), pp. 223, 229-230.

12. *The I-li or Book of Etiquette and Ceremonial,* tr. John C. Steele (London, 1917), pp. 189-242; *Li-chi,* tr. James Legge, editorial matter by Ch'u Chai and Winberg Chai (New Hyde Park, 1967), 2:458-464.

13. Creel, pp. 205-208, 211-212, and the *Tso-chuan* passages cited there.

14. Legge, 5: 424, cited by Ying-shih Yü, *Trade and Expansion in Han China* (Berkeley and Los Angeles, 1967), p. 5.

15. Yü, pp. 9-12; Ssu-ma Ch'ien, *Shih-chi,* chüan 110, in Burton Watson, tr., *Records of the Grand Historian* (New York, 1961), 2:155-192; Pan Ku, *Ch'ien Han-shu* 48:12b-13b.

16. *Shih-chi,* chüan 113, 114, 123, translated in Watson, 2:239-250, 258-289.

17. Pan Ku, chüan 8 in H.H. Dubs, tr., *History of the Former Han Dynasty,* 2: 258-259; Ssu-ma Kung, *Tzu-chih t'ung-chien* (reprint, Hong Kong, 1956), 27:885-888.

18. Pan Ku, Dubs tr., 3:120, 295-296, 304.

19. *Hou Han-shu* (po-na-pen, ed.), *chih,* 5:14-16; *chuan,* 79:5b.

20. *Ta-Ming chi-li,* 31:1-7b; *Hsin T'ang-shu* (Po-na-pen ed.), 16:1-2b.

21. Colin Mackerras, ed. and tr., *The Uighur Empire According to the T'ang Dynastic Histories: A Study in Sino-Uighur Relations, 744-840* (Columbia, 1972); Larry W. Moses, "T'ang Tribute Relations with the Inner Asian Barbarian," in John C. Perry and Bardwell L. Smith, eds., *Essays on T'ang Society: The Interplay of Social, Political, and Economic Forces* (The Hague, 1976), pp. 61-90; C.P. Fitzgerald, *Son of Heaven: A Biography of Li Shih-min, Founder of the T'ang Dynasty* (Cambridge, 1933), Chapter 7.

22. Mackerras, pp. 20, 72-74, 78, 80, 86, 95, 98-100, 116.

23. *Sung-shih* (Po-na-pen ed.), 119:9-22. Morris Rossabi, ed., *China Among Equals: The Middle Kingdom and Its Neighbors, 10th-14th Centuries* (Berkeley, 1983) appeared as I was doing final copy-editing for this book. It has saved me from a few errors and given me a few new insights, but I have not had time to take into account the wide range of sophisticated approaches to Sung foreign relations it opens up.

24. Christian Schwarz-Schilling, *Der Friede von Shan-yüan (1005 n. Chr.): Ein Beitrag zur Geschichte der chinesischen Diplomatie* (Wiesbaden, 1959), especially pp. 90-91.

25. James T.C. Liu, *Ou-yang Hsiu: An Eleventh-Century Neo-Confucianist* (Stanford, 1967); note also the greater tendency to fit accounts of relations with the Uighurs into a framework of Chinese superiority in the *Hsin T'ang-shu,* while the same events are more frankly described in *Chiu T'ang-shu;* Mackerras, pp. 64-65, 86-87.

26. *Sung-shih,* 485:23-23b.

27. Dagmar Thiele, *Der Abschluss eines Vertrages: Diplomatie zwischen Sung-und Chin-Dynastie, 1117-1123* (Wiesbaden, 1971).

28. *Sung-shih,* 119:17-17b.

29. Helmut Wilhelm, "From Myth to Myth: The Case of Yueh Fei's Biography," in Arthur F. Wright and Denis Twitchett, eds., *Confucian Personalities* (Stanford, 1962), pp. 146-161.

30. Yun-yi Ho, "Ritual Aspects of the Founding of the Ming Dynasty," *Bulletin of the Society for the Study of Chinese Religions,* 7:58-70 (Fall 1979).

31. Meng Sen, *Ming-tai shih* (Taipei, 1957), 27.

32. Tilemann Grimm, *Erziehung und Politik im konfuzianischen China der Ming-Zeit, 1368-1644* (Hamburg, 1960).

33. *Ming hui-tien* (Taipei reprint, 1968), 58:1439-1443.

34. *Ming hui-tien,* 58:1443-1447.

35. *Ming hui-tien,* 105:2286, 108:2337.

36. Wills, *Pepper,* pp. 4-7, and the works cited there.

37. Wang Gungwu, "The Opening of Relations Between China and Malacca, 1403-1405," in John Bastin and R. Roolvink, eds., *Malayan and Indonesian Studies: Essays Presented to Sir Richard Winstedt* (Oxford, 1964), pp. 87-104.

38. *Ta Ming chi-li,* 31:5b-6, 21b.

39. L. Carrington Goodrich and Chaoying Fang, *Dictionary of Ming Biography* (New York, 1976), pp. 144-145, 194-200, 360-362. There is a large and still-growing literature on the Cheng Ho maritime expeditions, which soon will include an important posthumous essay by Lo Jung-pang.

40. Wang Yi-t'ung, *Official Relations Between China and Japan, 1368-1549* (Cambridge, Mass., 1953), Ch. 3.

41. Wang Gungwu.

42. Goodrich and Fang, p. 959.

43. *Ming hui-tien,* 105:2291; Lung Wen-pin, *Ming hui-yao* (Peking reprint, 1956), 78:1515.

44. Serruys, *Tribute System,* Chapter 7; *Ming hui-tien,* 108:2321.

45. Lung Wen-pin, 80:1555; Fletcher, "China and Central Asia," in Fairbank, *Chinese World Order,* pp. 208-209, 346-347.

46. Wills, *Pepper,* pp. 7-8; Wang Yi-t'ung, pp. 76-77.

47. Serruys, *Tribute System,* Chapter 4.

48. Henry Serruys, C.I.C.M., *Sino-Jurčed Relations during the Yung-lo Period, 1403-1424* (Wiesbaden, 1955), p. 23.

49. Morris Rossabi, *China and Inner Asia: From 1368 to the Present Day* (London, 1975), pp. 44-45.

50. Serruys, *Tribute System,* pp. 34, 52-76.

51. Ibid., pp. 114, 354-359, 379-381.

52. Goodrich and Fang, pp. 1029-1035.

53. Morris Rossabi, "Ming China and Turfan, 1406-1517," *Central Asiatic Journal,* 16.3:206-225 (1972).

54. Serruys, *Tribute System,* pp. 589-605.

55. John E. Wills, Jr., "Maritime China from Wang Chih to Shih Lang: Themes in Peripheral History," in Jonathan D. Spence and John E. Wills, Jr., eds., *From Ming to Ch'ing: Conquest, Region, and Continuity in Seventeenth-Century China* (New Haven and London, 1979), pp. 211-215.

56. Wills, *Pepper,* p. 9. The best collection of Chinese documents on Ming relations with Europeans is Chang Wei-hua, *Ming-shih Fo-lang-chi, Lü-sung, Ho-lan, I-ta-li-ya ssu-chuan chu-shih* (Peiping, 1934).

57. Wills, "Maritime China," p. 8.

58. One of the few exceptions is a series of changes on the Ryukyus; Lung Wen-pin, 77:1504-1505.

59. T. C. Lin, "Manchuria in the Ming Empire," *Nankai Social and Economic Quarterly,* 8.1:1-43 (1935); Morris Rossabi, *The Jurchens in the Yuan and Ming* (Ithaca, 1982).

60. Serruys, *Tribute System,* pp. 57-72; Wills, "Maritime," pp. 215, 220.

61. Wills, *Pepper,* p. 13.

62. *Ta-Ch'ing t'ung-li,* 40:73-74, 45:1-6b; *HTSL,* 505:1-4, 510:1-2b, 510:16b-17b, 511:1-2b, 514:1-2b, 519:1-2; *LPTL,* 171:1-15, 185:1-2, 199:2b; *Ta-Ch'ing hui-tien,* K'ang-hsi, 72:1-3b, 73:12-17, 74:5; *Ch'ing-ch'ao t'ung-chih,* 46:7019-7022. Comparison of all these texts, and of intervening editions of *Ta-Ch'ing hui-tien shih-li* and *Ta-Ch'ing hui-tien tse-li* shows that their editors all had access to the same files of documents, and all made small changes and omissions in copying them.

63. See also Serruys, *Tribute System,* p. 344.

64. This was reported by Vincent Paats; see Chapter 5.

65. Wills, *Pepper,* pp. 110-112.

66. See Chapter 5, Note 12.

67. J. K. Fairbank and S. Y. Teng, "On the Transmission of Ch'ing Documents," in Fairbank and Teng, *Ch'ing Administration: Three Studies* (Cambridge, Mass., 1960), pp. 1-35, at 24-25.

68. Personal information and photographs from Mr. Herman Wong, 1979.

69. See Chapter 4, Note 26 below.

70. *HTSL,* ch. 505, cites many cases of such favors.

71. F. Pimentel, p. 22, translated in Appendix A.

72. See, for example, Serruys, *Tribute System,* p. 73, on Altan bowing toward the south in the 1571 ceremonies.

73. Jonathan Spence, "Ch'ing," in K. C. Chang, ed., *Food in Chinese Culture: Historical and Anthropological Perspectives* (New Haven and London, 1977), p. 283; *HTSL,* 1089:3b-4b, 17b-18b, 23.

74. Ho Ch'ang-ling, ed., *Huang-ch'ao ching-shih wen-pien* (Taipei reprint, 1966).

75. The highest is 6 of 38 in *Ta-Ch'ing hui-tien,* K'ang-hsi; others include *Ta-Ming hui-tien,* 8 of 73; *LPTL,* 15.5 of 151; *HTSL,* 14 of 185.

TWO: PIETER VAN HOORN, 1666-1668

1. Wills, *Pepper,* pp. 29-135.

2. *SL:SC,* 32:18-18b; *Ch'ing-ch'ao wen-hsien t'ung-k'ao,* 297:7461, 7464; *HTSL* 502:3-4; *LPTL* 173:1; Chang Wei-hua, pp. 104-105.

3. See Chapter 3.

4. For a general picture of this world, see my "Maritime China"; for a little more detail on the Dutch role in it, see my "De V.O.C. en de Chinezen in China, Taiwan en Batavia in de 17de en de 18de Eeuw," in M.A.P. Meilink-Roelofsz, ed., *De V.O.C. in Azië* (Bussum, 1976), pp. 157-192.

5. W. Ph. Coolhaas, ed., *Generale Missiven van Gouverneurs-Generaal en Raden aan Heren XVII der Verenigde Oostindische Compagnie,* 5 vols. to date (The Hague, 1960, 1964, 1968, 1971, 1975), 2: 394, 452, 538.

6. Ibid., 2:606; L. Pfister, S.J., *Notices Biographiques et Bibliographiques sur les Jésuites de l'Ancienne Mission de Chine (1552-1773)* 2 vols. (Variétés Sinologiques, Nos. 59, 60; Shanghai, 1932, 1934), pp. 256-263.

7. Johan Nieuhoff, *Het Gezantschap der Neerlandsche Oost-Indische Compagnie aan den Grooten Tartarischen Cham, Den Tegenwordigen Keizer van China* (Amsterdam, 1665), pp. 21-26; *Dagh-register gehouden in 't Casteel Batavia, 1628-1682* (Batavia, 1887-1931), 1653, pp. 51-64, 113-114, 159-165; Lo-shu Fu, *Documentary Chronicle,* pp. 11-12; *Ming-Ch'ing shih-liao,* 3d collection, 336-338b.

8. *SL:SC,* 68:2-3.

9. Nieuhoff; Fu, *Documentary Chronicle,* pp. 16-20. This embassy awaits study from the rich Dutch manuscript sources.

10. Wills, *Pepper,* pp. 108-109.

11. *SL:SC,* 102:10; Fu, *Documentary Chronicle,* pp. 20-21.

12. Wills, *Pepper,* pp. 25-28. The current Dutch project for the publication of the day-registers of Casteel Zeelandia should immensely facilitate further study of this topic.

13. Wills, *Pepper,* Chapters 2, 3.

14. Viraphol, *Tribute and Profit,* pp. 31-32.

15. Ibid., p. 32.

16. Wills, *Pepper,* p. 135.

17. VOC 680:387-388, 681:58, 77-78, 85-88, 99-100, 101-104, 112-113, Batavia resolutions 12 January-25 June 1666; VOC 890:295-327, Batavia credentials and instructions for Van Hoorn and letters to the Emperor and the Fukien officials, 2 July 1666.

18. Johan E. Elias, *De Vroedschap van Amsterdam, 1578-1695* (Haarlem, 1903-1905), I:167-168, 270, 295-298, 373-374, 447, 544-545.

19. M. A. van Rhede van der Kloot, *De Gouverneurs-Generaal en Commissarissen Generaal van Nederlandsch-Indië* (The Hague, 1891), pp. 72-75.

20. VOC 890:296-298.

21. Wills, *Pepper,* pp. 182-183.

22. A. J. Bernet Kempers, ed., *Journaal van Dircq van Adrichem's Hofreis naar den Groot-Mogul Aurangzeb, 1662* (Works of the Linschoten Vereeniging, XL, The Hague, 1941), pp. 46, 160; J. Feenstra Kuiper, *Japan en de Buitenwereld in de Achttiende Eeuw* (The Hague, 1921), pp. 179-80.

23. VOC 1253:1848-1883, Foochow to Batavia, 31 October 1665, at 1863-1864.

24. Apparently no full Dutch list of these presents has been preserved. For Chinese lists, see *HTSL,* 503:4b-5, *LPTL,* 178:2, Liang T'ing-nan,

Yueh hai-kuan chih, 22:10b-11b. For a Portuguese translation from the Chinese lists, see Ajuda/JA 49-V-15:338-347v.

25. Wills, *Pepper,* Chapter 2, especially p. 39 and map on p. 102.

26. For the embassy's first stay in Foochow, see VOC 1258:1391-1486 and VOC 1264:123-a171, Embassy resolutions, day-register, and letters, 24 August 1666-22 February 1667; VOC 1257:1366-1383, Embassy to Batavia, 14 July and 8 November 1666; VOC 1258:1384-1390, P. van Hoorn to Batavia and to the Amsterdam Chamber of the VOC, 6 November 1666. Many of the day-registers of the embassy are reproduced with some changes and omissions in Olfert Dapper, *Gedenkwaerdig Bedryf der Nederlandsche Oost-Indische Maetschappye op de Kuste en in het Keizzerrijk van Taising of Sina,* pp. 211-397. The most important omissions are of the texts of some translations of Chinese documents. An English translation of this work by John Ogilby was published in London in 1671 as *Atlas Chinensis . . . by Arnoldus Montanus.* Ogilby's English translation contains some materials not in Dapper, so Ogilby probably worked from a different manuscript copy. Dapper contains at least one short passage skipped in the VOC manuscript. There is still another copy of these day-registers in Algemeen Rijksarchief, Van Hoorn-Van Riebeeck Collection, No. 5, that contains a very small amount of unimportant material not in any of the other texts. For more on Dapper, see Wills, *Pepper,* p. 18, Note 39. On Captain Carvalho, see Wills, "The Hazardous Missions of a Dominican: Victorio Riccio, O.P., in Amoy, Taiwan, and Manila," in *Actes du IIe Colloque International de Sinologie, Chantilly* (Paris, 1980), pp. 231-257, at 254, citing *inter alia* Arsip Nasional Republik Indonesia, Gereformeerde Kerk 63, Brieven van 1627-1737, fol. 141-141v, P. van Hoorn to Batavia Church Council, 8 November 1666. On Ch'i-ying's interest in Pottinger's family, see John K. Fairbank, *Trade and Diplomacy on the China Coast* (Cambridge, 1953), I:110-112.

27. VOC 1265:1105-1145, Day-register of party in Foochow, 25 February-24 October 1667.

28. Robert B. Oxnam, *Ruling from Horseback: Manchu Politics in the Oboi Regency, 1661-1669* (Chicago and London, 1975), pp. 174-175.

29. VOC 1265:1114v-1118v, Foochow day-register, 4-9 May 1667. Gabriel de Magalhaens, S.J., also reported that this was the reason for the punishment of the officials; ARSJ/JS 124:55-57v, G. de Magalhaens, S.J., to Luis da Gama, S.J., 23 and 25 April 1667, at 57v; Adrien Greslon, S.J., *Histoire de la Chine sous la Domination des Tartares* (Paris, 1671), pp. 307-308. I have found no mention of Chang's dismissal in any Chinese source; in the standard biographical collections, his career simply ends abruptly and inexplicably, or it is stated that he requested retirement because of age and ill health. On the Dowager Empress Hsiao-chuang, see

Silas H. L. Wu, *Passage to Power: K'ang-hsi and his Heir Apparent, 1661-1772* (Cambridge, Mass. and London, 1979), Chapter 1.

30. Wills, *Pepper*, p. 140; Chapter 3 below.

31. VOC 1265:1037-1070, Embassy day-register, 20 January-20 June 1667.

32. On these flash-locks, see Joseph Needham, *Science and Civilization in China* (Cambridge, 1954-) Vol. IV, Part 3, pp. 215, 347; Gabriel de Magalhaens, S.J., *Nouvelle Relation de la Chine* (Paris, 1688), p. 141.

33. On the stay in Peking, see VOC 1265:1016-1036, Embassy to Batavia, 13 November 1667; VOC 1265:1071-1093, Embassy day-register 20 June-5 August 1667; ARSJ/JS 124:71-80v, "Relacion del Estado de la Christainidad de Gran Imperio de China", anon., 1669, at 78v-79; ARSJ/JS 162:176-181v, L. Buglio, S.J. to Feliciano Pacheco, S.J., 13 July 1667, at 179-180 (a very well-informed summary); *SL:KH* 22:10; *Ta-Ch'ing hui-tien*, K'ang-hsi, 72:13a-b, 74:8b-9b, 77:20b-21; *LPTL*, ch. 178; *HTSL*, 503:4b-5, 506:5b, 510:17, 511:2b; Chang Ju-lin and Yin Kuang-jen, *Ao-men chi-lueh*, in *Chao-tai ts'ung-shu*, 35b; *Ch'ing-ch'ao wen-hsien t'ung-k'ao*, 298:7473; Liang T'ing-nan, *Yueh hai-kuan chih*, 22:6-7, 10b-11; Liang T'ing-nan, *Yueh-tao kung-kuo shuo*, 3:11-14b, the last embodying some confusions with the Paats embassy.

34. Oxnam, pp. 182-198.

35. VOC 1265:1131v-1133, Day-register of party in Foochow, 14 August 1667.

36. Oxnam, pp. 185-198.

37. VOC 1269:294v-298v, Translations of imperial edicts, etc.; *Ta-Ch'ing hui-tien*, K'ang-hsi, 72:13; *HTSL*, 510:17. Limitation of maritime trade to that in connection with tribute embassies was confirmed and made general on 19 May 1668; *SL:KH*, 25:22.

38. The basic sources for the rest of the embassy are VOC 1265:1094-1104, Embassy day-register, 5 August-2 November 1667; VOC 1267:620-638, Embassy day-register, 2 November 1667-9 January 1668; VOC 1267: 604-619, Van Hoorn's report at Batavia, 10 March 1668; VOC 1269:273-308v, another copy of Van Hoorn's report and related documents. Adrien Greslon, S.J., in his *Histoire de la Chine sous la Domination des Tartares*, pp. 307-308, reports that, on their way south, the Dutch were forbidden to buy or accept anything except rice, and that the governors of the provinces they passed through had orders to count the Dutchmen!

39. VOC 682:63-64, Batavia resolution, 13 May 1667; VOC 891:290-318, Batavia to Foochow, 13 June 1667.

40. *Dagh-register*, 1668, pp. 2-3, 9-16.

41. VOC 1269:299-301v, Discussion by Van Hoorn, n.d., spring 1668.

42. F. de Haan, *Priangan* (Batavia, 1910-1912) Vol. I, "Personalia", p. 4. For a dispute between Van Hoorn and the Batavia Council on the finances of his embassy, see VOC 683:74-75, 78, 89, 90-96, 121-122, 149, 155,

and one not numbered, Batavia resolutions 13 April 1668-29 January 1669. For figures on expenses and trade profits of the embassy, see Pieter van Dam, *Beschryvinge van de Oostindische Compagnie* (The Hague, 1927-1954), Book II, Part I, p. 740.

43. Henri Cordier, *Bibliotheca Sinica* (Paris, 1906-1907), columns 2347-2350.

44. Pieter van Hoorn, *Eenige Voorname Eygenschappen van de Ware Deugdt, Wysheydt, en Volmaecktheydt, Getrocken uyt den Chineschen Confucius* (Batavia, 1675). I am deeply indebted to Katharine S. Diehl for sending me a photocopy of this extremely rare book.

45. Wang Shih-chen, *Ch'ih-pei ou-t'an,* reprinted in *Pi-chi hsiao-shuo ts'ung-shu* (Shanghai, 1935), 2:194-195; Wang Shih-chen, *Yü-yang ching-hua lu* (1700; Taipei reprint, 1966), 6:12-12b (For a mis-translation of the last line of the sword poem, see my *Pepper,* p. 105); Ch'en Wei-sung *Hu-hai-lou shih-chi* (1689), 2:26b-27. Another poem that may be about this embassy is Lao Chih-pien, *Ching-kuan-t'ang shih-chi,* 2:1. Another poem about the Dutch from this period is a purely antiquarian effort by a scholar of late Ming foreign relations; Yu T'ung, "Wai-kuo chu-chih tz'u," pp. 11b-12; in his *Yu Hsi-t'ang chi.* I am grateful to Lo-shu Fu for references to several of these poems.

THREE: MANOEL DE SALDANHA, 1667-1670

1. Important modern studies of this embassy are Lo-shu Fu, "The Two Portuguese Embassies to China during the K'ang-hsi Period," *T'oung Pao,* 43:75-94 (1955), at 75-87; Luciano Petech, "Some Remarks on the Portuguese Embassies to China in the K'ang-hsi Period," *T'oung Pao,* 44: 227-241 (1956), at 227-233, and Fu, *Documentary Chronicle,* pp. 41, 46-47, 450, 453. Others include Eduardo Brazão, "The embassy of Manuel de Saldanha to China in 1667-1670 (Notes on Sino-Portuguese diplomatic relations)," *Boletim do Instituto Portuguès de Hongkong,* 1:139-162 (July 1948). Brazão, *Apontamentos para a Historia das Relações Diplomáticas de Portugal com a China, 1516-1753* (Lisbon, 1949), pp. 67-107, also published as a separate pamphlet, *Subsídios para a Historia da Relações Diplomáticas de Portugal com a China: A Embaixada de Manuel de Saldanha, 1667-1670* (Macao, 1948); Paul Pelliot, "L'ambassade de Manoel de Saldanha a Pekin," *T'oung Pao,* 27:421-424 (1930); Durval R. Pires de Lima, *A Embaixada de Manuel de Saldanha ao Imperador K'hang hi em 1667-1670* (Lisbon, 1930); Chou Ching-lien, *Chung-P'u wai-chiao shih* (Shanghai, 1936), pp. 155-157; Antonio da Silva Rego, "Macau entre duas crises (1640-1688)," *Anais da Academia Portuguesa de História,* 24.2:307-334 (1977), 323-325; C. A. Montalto de Jesus, *Historic Macao* (Shanghai, 1902), pp. 101-104; Andrew Ljungstedt, *An Historical Sketch*

of the Portuguese Settlements in China (Boston, 1836), pp. 95-96. Secondhand contemporary references in Western languages include H. Bosmans, S.J., *Lettres Inédites de Francois de Rougemont* (Louvain, 1913), 29; Joseph Sebes, S.J., *The Jesuits and the Sino-Russian Treaty of Nerchinsk (1689): The Diary of Thomas Pereira, S.J.* (Bibliotheca Instituti Historici S.I., Vol. 18, Rome, 1961), pp. 208-211; Ferdinand Verbiest, S.J., *Correspondance*, ed. H. Josson and L. Willaert, (Brussels, 1938), pp. 260-262, 275-276, 334-335; Ajuda/JA 49-IV-62:1-10 and 118-127v, Annual Letter of the Japan Province, probably by Luis da Gama, S.J., 18 November 1668; ARSJ/JS 104:194v-195, 200-201v, 241-243, passages in T. I. Dunin Szpot (see Note 16 below); ARSJ/JS 122:327-363, A. Greslon, S.J., to Father General, 20 October 1670, at 356-356v, 360-362; ARSJ/JS 124: 71-80v, "Relacion del Estado de la Christianidad de gran Imperio de la China," at 79; ARSJ/JS 124:81-86v, "Novas da China, Anno 1669," at 85-85v; ARSJ/JS 124:87-88v, Brother Miguel dos Anjos, S.J., 1 February 1670. Chinese bureaucratic sources are *Ta Ch'ing hui-tien*, K'ang-hsi, 72: 18-19b, 74:10b, 77:21b-22; *HTSL*, 502:20b, 503:5, 506:6, 510:2-2b, 514:1b-2, 520:8-8b; *LPTL*, 180:1, 2, 4b-5; (180:10 seems to be a misfiled reference to one of the eighteenth-century Portuguese embassies); *SL:KH*, 33:20, 34:10; *Ch'ing-ch'ao wen-hsien t'ung-k'ao*, 298:7468; Liang T'ing-nan, *Yueh hai-kuan chih*, 22:19v; Liang T'ing-nan, *Yueh-tao kung-kuo shuo*, 4:4b-8; Chang Ju-lin and Yin Kuang-jen, 23b, 41; Wang Shih-chen, *Ch'ih-pei ou-t'an*, 1:3.

2. C. R. Boxer, *Fidalgos in the Far East* (1948; reprint, London, 1968), p. 121; Benjamin Videira Pires, S.J., *Embaixada Mártir* (Macao, 1965).

3. C. R. Boxer, *Macau na Época da Restauração* (Macao, 1942), pp. 153-155.

4. Boxer, *Fidalgos*, pp. 147-150; Boxer, *Restauração*, p. 45.

5. Hsu Tzu, *Hsiao-t'ien chi-nien* (*TW*, No. 168), 648-649, 810-811; Ajuda/JA 49-IV-61:75-131, Annual letter for 1651, at 91-121v; Ajuda/JA 49-IV-61:252v-260, "Account of what occurred in the siege of Canton by the Tartars," n.d.

6. Antonio Francisco Cardim, S.J., *Batalhas da Companhia de Jesus na sua Gloriosa Provincia de Japão* (Lisbon, 1894), pp. 21, 29-30; Boxer, *Fidalgos*, p. 144, cites an estimate of 40,000 in Macao in 1644.

7. C. R. Boxer, "Portuguese Military Expeditions in Aid of the Mings against the Manchus, 1621-1647," *T'ien-hsia Monthly*, 7.1:24-50 (August 1938).

8. C. R. Boxer, *Francisco Vieira de Figueiredo: A Portuguese Merchant-Adventurer in South East Asia, 1624-1667* (Verhandelingen van het Koninklijk Instituut voor Taal-, Land-, en Volkenkunde, No. 52, The Hague, 1967).

9. C. R. Boxer, "The Rise and Fall of Nicholas Iquan," *T'ien-hsia Monthly*, 11.5:401-439 (April-May 1939), p. 429.

10. *SL:SC*, 33:18b; *Ming-Ch'ing shih-liao*, Collection 3, No. 4, p. 307; Lo-shu Fu, *Documentary Chronicle*, pp. 8, 9; *Ch'ing-ch'ao wen-hsien t'ung-k'ao*, 298:7473.

11. I hope to publish the results of my own research on the origins of Cheng Chih-lung in the near future.

12. Arthur W. Hummel, ed., *Eminent Chinese of the Ch'ing Period* (Washington, 1943), pp. 634-636, 792-798; ARSJ/JS 162:196-197v, L. da Gama, S.J., to Father General, 25 October 1667, at 196, ARSJ/JS 162: 241-241v, Da Gama to Father General, 10 December 1668.

13. See Chapter 2 on the Dutch contacts, and Boxer, *Restauração*, p. 62.

14. All these appear in the diary of Luis da Gama, S.J.; see Note 47 to this chapter. Li Chih-feng seems to have remained influential until the 1680s; scattered evidence on his career and those of several other merchants is pulled together in my work in progress on the 1680s.

15. Wills, *Pepper*, p. 44.

16. Ajuda/JA 49-V-3:199-201v, Report on insults and disorders by Chinese in Macao, 17 August 1658; ARSJ/JS 104, Thomas Ignatius Dunin Szpot, "Collectanea Historiae Sinensis, 1641-1707," at 143v-146. This important manuscript work summarizes a large number of Jesuit reports from China, including many that survive in no other form.

17. Wills, "Early Sino-European Relations," pp. 67-68, 73.

18. Boxer, *Fidalgos*, pp. 153-155; C. R. Boxer, *Portuguese Society in the Tropics: The Municipal Councils of Goa, Macao, Bahia, and Luanda, 1510-1800* (Madison and Milwaukee, 1965), Chapter 2.

19. Hsieh Kuo-chen, "Ch'ing-ch'u tung-nan yen-hai ch'ien-chieh k'ao, *Kuo-hsueh chi-k'an*, 2.4:797-826 (December 1930); English translation in *Chinese Social and Political Science Review*, 15:559-596 (1931).

20. Tanaka Katsumi, "Shinsho no Shina enkai: Senkai o chūsin to shite mitaru," *Rekishigaku kenkyū*, 6.1:73-81, 6.3:83-94 (1936), at 79; *Ch'ing ch'i-hsien lei-cheng hsuan-pien* (TW, No. 230), 413.

21. Wills, *Pepper*, pp. 53-57, 66-68, 127, 139-141.

22. Goa, Monções 35:43-51, Viceroy to King, n.d. This excellent letter is the source of the rest of our information on 1662, unless otherwise noted.

23. Wills, *Pepper*, p. 55.

24. Boxer, *Francisco Vieira de Figueiredo*, pp. 84-85.

25. The Goa letter states the population of Portuguese and other Christian males was 302; VOC 1244:2185-2228, Day-register in Foochow, 28 February-2 September 1663, at 2186v-a2188v, 22-23 May, reports a very interesting conversation with a deserter from the Dutch forces in Taiwan who had been in Macao in 1662; he estimated 200 males and 2,000 widows and orphans.

26. The quotation is from the Goa letter cited in Note 22 above.

27. Boxer, *Francisco Vieira de Figueiredo*, p. 85.

28. ARSJ/JS 104:156, Dunin Szpot.

29. François de Rougemont, S.J., *Historia Tartaro-Sinica Nova* (Louvain, 1673), p. 73.

30. AHU, Macau, maço 1, Senate of Macao to Viceroy, 14 November 1666.

31. Goa letter, as cited.

32. Goa letter, as cited. We have more information from the Dutch deserter mentioned in Note 25 above, some of it a little hard to reconcile with other sources. He said that Chinese merchants were allowed to sell daily necessities in Macao, but were so much vexed by Ch'ing officials that they said they sooner or later would turn on their oppressors. He also reported that the Macao authorities had had to pay 13,000 taels to Shang K'o-hsi to get a merchant and some priests out of prison.

33. Goa letter as cited; ARSJ/JS 104:156, Dunin Szpot; Ajuda/JA 49-IV-62: 1-10, Annual letter of Japan Province of Society of Jesus, 1668, at 2-3.

34. See Chapter 2 on the revocation of the Dutch privilege, and F. Pimentel, S.J., *Breve Relação da Jornada que Fez à Corte de Pekim o Senhor Manoel de Saldanha, Embaixador Extraordinario del Rey de Portugal ao Emperador da China, e Tartaria (1667-1670)*, ed. C. R. Boxer and J. M. Braga (Macao, 1942), p. 38. Pimentel's text, pp. 11-45 of the Boxer-Braga volume, is translated in Appendix A of this book. Footnotes here refer to pages in the Boxer-Braga volume; my translation is keyed to these pages. The Boxer-Braga volume also contains texts of other documents related to the Saldanha embassy.

35. Lo-shu Fu, *Documentary Chronicle*, pp. 35, 447.

36. Rougemont, pp. 72-83; Philippe Chahu, S.J., *Lettres des Pays Estrangers* (Paris, 1668), pp. 122-124; *Dagh-register Gehouden in 't Casteel Batavia, 1628-1682*, 1663, p. 98.

37. Lo-shu Fu, *Documentary Chronicle*, 38; Wills, *Pepper*, pp. 118-119.

38. Lo-shu Fu, *Documentary Chronicle*, pp. 35-38; Domingo Fernandez Navarrete, O.P., *Travels and Controversies of Friar Domingo Navarrete, 1618-1686*, ed. J. S. Cummins (London, 1962), pp. 246-249; Rougemont, pp. 188-327.

39. Wills, *Pepper*, pp. 115-116.

40. AHU, Macau, maço 1, Macao Senate to Viceroy, 17 December 1664; Wei Yuan, "Kuo-ch'u tung-nan ching-hai chi," in *Hai-pin ta-shih chi* (TW, No. 213), pp. 37-44, at 43.

41. VOC 1244:a2188; see Note 25 above.

42. AHU, Macau, maço 1, Macao Senate to Viceroy, 17 December 1664 and 14 November 1666.

43. This is according to the Chinese documents cited in Note 63 below. Could this have been the same transaction for 13,000 taels mentioned in Note 32 above?

44. See the letter of this date cited in Note 42 above.

45. This is the approximate number of ships on hand mentioned in the opening entries of the Da Gama diary; see Note 47 below.

46. Da Gama diary, p. 36.

47. J. F. Marques Pereira, "Uma resurreição histórica (Páginas inéditas dum visitador dos jesuitas, 1665-1671)," *Ta-Ssi-Ynag-Kuo* (Macao), I:31-41, 113-119, 181-188, 305-310; II: 693-702, 747-763 (1899 and 1901). Unless otherwise noted this is the source for all that follows up to January 1670.

48. Identifications from *Kuang-tung t'ung-chih* (Yung-cheng ed.), ch. 29.

49. The "mandarin of Ansão" in Da Gama.

50. Li Yuan-tu, *Ch'ing hsien-cheng shih-lueh hsuan* (*TW*, No. 194), pp. 1-6.

51. Marques Pereira, p. 36.

52. Nieuhoff, p. 166.

53. Marques Pereira, pp. 36-37.

54. Ibid., pp. 113-114.

55. AHU, Macau, maço 1, Senate of Macao to Viceroy, 15 November 1666.

56. Navarrete, p. 274. Navarrete gives an account of the travails of Macao and of the beginnings of the Saldanha embassy that is basically fairly accurate but full of touches of contempt for the Ambassador and the Jesuits; pp. 234, 269-273.

57. Marques Pereira, p. 115. At an earlier point in their negotiations, the Portuguese had understood that the Canton officials had told the court the goods on hand in Macao all had arrived there before the maritime trade prohibitions of 1662; actually it seems that all had arrived in 1664 or later.

58. Marques Pereira, p. 115.

59. Oxnam, *Ruling from Horseback*, pp. 170-175.

60. Marques Pereira, pp. 185, 305-306; Greslon, *Historie de la Chine*, pp. 305-314.

61. Chapter 2; Wills, *Pepper*, pp. 111-112.

62. Oxnam, pp. 185-186; Verbiest, *Correspondance*, p. 121.

63. Ajuda/JA 49-IV-62: 1-10, Annual letter for 1668, 18 November 1668, at 3-4. For Chinese documentation on the charges against Lu Hsing-tsu and others, see *Cheng-shih shih-liao san-pien* (*TW*, No. 175), pp. 74-83. There is general agreement between these documents and the Da Gama diary, but it is very hard to reconcile them in detail. For reports of later memorials against Macao, see Ajuda/JA 49-IV-62; 443, 449, Running summary. On Wang Lai-jen, see Lo-shu Fu, *Documentary Chronicle*, pp. 40-41.

64. Panduronga S. S. Pissurlencar, ed., *Assentos do Conselho do Estado da India* (Bastorá, 1953-1958), 4:178, 183; Goa, Assentos do Conselho da Fazenda, 11:108v-109, 27 April 1667; Goa, Cartas Patentes e Alvaras,

44:36v-37, 29 April 1667. I am most grateful to Father Teotonio R. de Souza, S.J., for sending me these references to manuscript sources I did not find time to consult in Goa. In several letters preserved in a letter-book of the embassy (Goa 1210), Saldanha claimed the royal power of appointment east of the Sunda Strait, but I doubt that he was able to make any effective use of that power.

65. Pissurlencar, 4:182.

66. Personal communication from George B. Souza.

67. AHU, Macau, maço 1, Administrators of Portuguese India to Prince Regent, 15 December 1667.

68. Luis de Menezes, Conde de Ericeira, *História de Portugal Restaurado* (Porto, 1945), 3:30-54; D. R. Pires de Lima, pp. 8-13; Goa, Monções 29-30:44, Viceroy to Prince Regent, 26 January 1666.

69. Personal communication from Robert J. Moes, M.D., June 1980.

70. Manuel Teixeira, "A Embaixada de Manuel de Saldanha a Pequim," *Boletim Ecclesiástico da Diocese de Macau,* No. 654 (October 1957), 891-894; Coolhaas, ed., *Generale Missiven,* III:573, 584.

71. AHU, Macau, maço 1, Copy of royal letter to Emperor of China, dated 12 March 1666; no doubt it had been given this date so that it would be plausible that it had come from Lisbon. Goa, Monções 36:44, Prince Regent to Viceroy, 6 February 1670.

72. VOC 1264:238-239, Malacca to Batavia, 22 August 1667.

73. F. Pimentel, p. 34, in Appendix A.

74. Marques Pereira, p. 699.

75. These letters, in the Ajuda/JA collection, are printed at the end of the Boxer-Braga edition of Pimentel, pp. 47-74. The accounts, from a manuscript in Macao, are in Pimentel, Appendix, pp. iii-xxvii. Also relevant to embassy financing are Goa 1210:22-24, Saldanha to Alvaro da Silva, 31 December 1668 and 1 January 1669, urging the City to pay its subsidies to the embassy promptly; Ajuda/JA 49-V-15:347v-349v and 352-357v, Records of expenses of embassy.

76. Goa 1210:22-24, Saldanha to Alvaro da Silva, 31 December 1668 and 1 January 1669; Pimentel, 57-59.

77. Goa 1210:36-37, Order establishing a committee to manage funds in Macao, 6 August 1669.

78. Pimentel, p. 34.

79. VOC 1272:1123-1224, Day-register of Canton voyage, 23 July 1668-7 May 1669, at 1128v and 1141v, 24 August and 26 September 1668. VOC 1272:1101-1122v, Report at Batavia on Canton Voyage, 7 May 1669, at 1114-1117v. The Dutch first were told that it had been doubted in Peking if the embassy had come from the King, but now it had been approved and would leave soon. Later they heard that the Ambassador still would not show the royal letter and was not allowed out of his lodgings, and that he was much hated because he was proud and stingy. Ajuda/JA 49-IV-

62:154-157, P. Buglio S.J. to L. da Gama S.J., 26 April 1668, at 154v, says letters from the Goa Viceroy to the Canton authorities had been forwarded to Peking, and the Peking Jesuits had been summoned to translate them in February 1668. See also Marques Pereira, p. 749.

80. Pimentel, p. 35; Ajuda/JA 49-IV-62:2. Early in December 1668, the Jesuits in Macao already had understood that the embassy had been accepted and would leave the following February or March; ARSJ/JS 162:241-241v, L. da Gama, S.J., 10 December 1668. The Peking Jesuits understood that some form of approval was sent from the capital about the end of November; ARSJ/JS 162:269-273, Peking Jesuits to Canton and Macao Jesuits, 2 January 1669, and Ajuda/JA 49-IV-62:408-412, L. Buglio S.J. to Macao Jesuits, 18 January 1669. Philippe Couplet, S.J., wrote to the Dutch on 17 February 1669 that the Canton officials had sent a report to Peking about Saldanha a month earlier than about the Dutch trading in the Canton area; the latter report probably was sent some time in October. Wills, *Pepper,* pp. 140-141; VOC 1272:1203v-1204, Day-register, Canton voyage, 25 February 1669. On Father Couplet's many contacts with the Dutch, see John E. Wills, Jr., "Some Dutch Sources on the Jesuit China Mission, 1662-1687," *Archivum Historicum Societatis Iesu,* in press. *Ta-Ch'ing hui-tien,* K'ang-hsi, 72:18, records that, in 1667, the arrival of an Ambassador and 4 ships was reported; in 1668, it was decreed that, in the future, only 3 ships with 100 men each might come with an embassy; and, in 1669, it was ordered that only 22 would be allowed to come to the capital and the rest of the party would be given rations.

81. Pimentel, pp. 54, 60-61; Marques Pereira, pp. 752-754.

82. Pimentel, pp. 49, 50, 56-59, 68-70; ARSJ/JS 104:201v, Dunin Szpot; Ajuda/JA 49-V-15:319-320, B. Pereira de Faria to Macao city authorities, 10 January 1670.

83. ARSJ/JS 162:241-241v, L. da Gama, S.J., 10 December 1668; Pimentel, pp. 43, 63-64; The Dutch party seeking trade in the Canton area saw the Portuguese on December 30 at a tent outside the city where they had successive interviews with Governor-General Chou; they understood that the Portuguese delegation had been headed by the second in command of the embassy, since the Ambassador was sick, and that they had simply turned in a list of the presents. VOC 1272:1160, Day-register, Canton voyage, 1 January 1669.

84. H. Bosmans, S.J., "Ferdinand Verbiest, Directeur de l'Observatoire de Peking (1623-1668)," *Revue des Questions Scientifiques,* 3rd series, 21: 195-273, 375-494, at 235-242; ARSJ/JS 162:269-273, G. De Magalhaens, S.J., to Jesuits in Canton and Macao, 2 January 1669.

85. Hummel, pp. 792-798; Pamela Crossley, "The Tong in Two Worlds: Cultural Identities in Liaodong and Nurgan during the 13th-17th Centuries," *Ch'ing-shih wen-t'i,* 4.9:21-46 (June 1983).

86. Hummel, pp. 663-66.

87. Ajuda/JA 49-IV-62:154-157, P. Buglio, S.J., to L. da Gama, S.J., 26 April 1668, at 154; ARSJ/JS 162:196-197v, L. da Gama, S.J., to Father General, 25 October 1667, at 196v.

88. Ajuda/JA 49-IV-62:420-460v, Running summary of letters from Peking Jesuits and events in Peking, 1669-1671, at 420v-421.

89. Pimentel, p. 14, 73:VOC 1272:1211v-1212v, 8 March 1669.

90. Saldanha complained of fevers and general ill health in a letter of 13 June 1669; Pimentel, p. 69.

91. Pimentel, p. 68.

92. Ajuda/JA 49-IV-62; 405v-408 and 421-424v, L. Buglio, S.J., to Macao Jesuits, 17 June 1669, and Running summary.

93. Pimentel, p. 71.

94. Ajuda/JA 49-IV-62; 543v-545, G. de Magalhaens, S.J., 31 October 1669; Pimentel, pp. 20, 21, 36; Ajuda/JA 49-IV-62:445v-446, Running summary.

95. To Peking and back the most important source is Pimentel; only points taken from other sources or out of chronological order in Pimentel are noted separately. Here see also Pimentel, p. 74, Marques Pereira, p. 756, Petech, p. 231.

96. According to Father Pimentel, some Chinese Christians sought him out on the boat simply because there was a cross on its flag.

97. Pimentel, pp. 14, 17, 36; ARSJ/JS 104:242v, Dunin Szpot; ARSJ/JS 162:345-349v, Tomasso Valguarnera, S.J., to Father General, 1 January 1673, at 345v.

98. Lung Wen-pin, 78:1521; Serruys, *Tribute System*, p. 160; *HTSL*, 503:10a-b.

99. Nieuhoff, pp. 55-158.

100. Pimentel, pp. 40-45; ARSJ/JS 104:200v, Dunin Szpot.

101. Pimentel, pp. 44-45.

102. Hummel, pp. 415-416.

103. Petech, p. 228.

104. Ibid.; Pimentel, p. 26, says it was the mansion of one of the four Regents; Ajuda/JA 49-IV-62:448-449 says it had belonged to "a count, an imperial relative." If it was the mansion of one of the Regents, Oboi seems most likely. Soni's family never fell from favor; Ebilun was back in the Emperor's graces in 1670; Suksaha is possible, but his fall was not so recent, and I think Jesuits would have made more of it because they saw him, not Oboi, as the great enemy. Hummel, pp. 219-221, 599-600, 663-664; Oxnam, Chapter 8.

105. Oxnam, pp. 189-191.

106. Chang Ju-lin and Yin Kuang-jen, in *Chao-tai ts'ung-shu*, 40.

107. Pimentel, p. 33.

108. ARSJ/JS 120:185-190, Summary of letter from Peking dated 11 August 1670; ARSJ/JS 104:241v, Dunin Szpot.

109. Pimentel, p. 17.

110. *SL:KH* does not record the date of his return.

111. Petech, p. 229; see Appendix D.

112. July 31 was the 15th day of the 6th month in the Chinese calendar. The *Veritable Records* make no mention of the embassy's audience.

113. Pimentel's list of 10 pieces of silk for the Ambassador, 6 for each of his three officers, and 4 for each of the other 9 Portuguese, (based on Pimentel's own statement that only 12 Portuguese accompanied the Ambassador) add up to 64 pieces, which is the figure given in letters by L. Buglio, S.J., and F. Marini, S.J., quoted by Petech, pp. 229, 231. For letters by Marini almost identical to that quoted by Petech, see ARSJ:JS 162:310-313, 320.

114. See Appendix D for the lists and sources. Wang Shih-chen, *Ch'ih-pei ou-t'an,* I:3, states with apparent reference to the Saldanha embassy that the Emperor ordered that the presents be increased by one degree or level (*chia i-teng*); this would seem to be more applicable to the bureaucratically regular increases bestowed on the Pereira de Faria embassy, and may suggest how the decisions on the two embassies became linked in the minds of some responsible officials.

115. Pimentel, pp. 21, 27, 36, 39.

116. Petech, p. 230.

117. Verbiest, *Correspondance,* p. 173-174. Verbiest also mentioned the problems of Macao in one of these conversations, but doubted that the Emperor understood.

118. *SL:KH,* 34:10. For translations of the imperial edict to the King of Portugal and the mourning edict for Saldanha, see Palha #3, and José de Jesus Maria, *Azia Sínica e Japónica,* ed. C. R. Boxer (Macao, 1941, 1950), pp. 80, 82. This work sometimes contains summaries of Macao documents that have disappeared, but, except for these documents, it has little to offer on 1662-1670 that is not more reliably available elsewhere. The translation of the edict to the King includes only part of the list of presents from the Emperor. For the regulations on the mourning for a tribute ambassador, see *HTSL,* 513:1-12b and *Ta-Ch'ing hui-tien,* K'ang-hsi, 72:2b-3.

119. Ajuda/JA 49-V-16:419-462v, Luis da Gama, S.J., "Just defense of the Society of Jesus. . . ."

120. Goa, Monções, 38B:304, Prince Regent to Viceroy, 19 September 1672.

121. Goa, Monções, 37:133-136, Viceroy to Prince Regent, 24 August 1672 and to Antonio Barbosa Lobo, 3 May 1672; Goa, Monções 39-40:4, Viceroy to Prince Regent, 4 December 1674.

122. *Ta-Ch'ing hui-tien,* K'ang-hsi, 72:18-18b; *LPTL,* 180:1.

FOUR: BENTO PEREIRA DE FARIA, 1678

1. There have been three modern studies of this embassy, all erudite and useful but none of them using all the available documentation: Lo-shu Fu, "Two Portuguese Embassies," pp. 87-92; Petech, pp. 233-236; and Giuliano Bertuccioli, "A Lion in Peking: Ludovico Buglio and the Embassy to China of Bento Pereira de Faria in 1678," *East and West* (Rome), New Series, 26.1-2:223-240 (1976). Other secondary mentions of it are in Lo-shu Fu, *Documentary Chronicle*, pp. 52, 456-457, and Chou Ching-lien, p. 57. Contemporary Western references include ARSJ/JS 104:305-307v, 104:338, and 105I:5-5b, Dunin Szpot; ARSJ/JS 124:137, Manoel Ferreira, S.J., 22 November 1668. Chinese bureaucratic sources are *SL:KH* 76:3-4, 77:7b, *HTSL* 503:5b, 506:6b, 510:2a-b; Liang T'ing-nan *Yueh hai-kuan chih*, 22:20; Liang T'ing-nan, *Yueh-tao kung-kuo shuo*, 4:6b-8; *Ch'ing-ch'ao wen-hsien t'ung-k'ao*, 298:7468; Chang Ju-lin and Yin Kuang-jen, 23b-24.

2. Marques Pereira, pp. 753-755.

3. Goa, Monções 36:282-287, Memorandum of the complaints that come from Macao . . . against D. Alvaro da Silva; Monções 37:133-136, Viceroy to Prince Regent, 24 August 1672 and to Antonio Barbosa Lobo, 3 May 1672.

4. VOC 686:91-92, 97-103, Batavia Resolutions, 26 and 29 May 1671.

5. Wills, *Pepper*, pp. 151-152; *Dagh-Register*, 1672, p. 80.

6. *Dagh-Register*, 1673, pp. 1-2, 319-320; 1674, p. 354; 1675, pp. 53-76.

7. José de Jesus Maria, pp. 89-90; Goa, Monções 37:133-136, as cited in Chapter 3, Note 121.

8. AHU, Macau, maço 1, Senate of Macao to Prince-Regent, 10 January 1673; VOC 1320:313-330v, Lampacao to Batavia, 29 October 1676, at 329-329v.

9. Wills, *Pepper*, pp. 154, 157-160.

10. Ibid., 153-154.

11. VOC 1320:329v-330, Lampacao to Batavia, 29 October 1676.

12. Bosmans, "Verbiest . . . Directeur," pp. 389-398, 405.

13. Jonathan D. Spence, *Emperor of China: Self-Portrait of K'ang-hsi* (New York, 1974), pp. 9, 11, 18-22, 65-75.

14. Goa, Monçoes 39-40:4, Viceroy to Prince-Regent, 4 December 1674; Goa, 1264:16-16v, Administrators of Portuguese India to City of Macao, 7 May 1678; VOC 1320:332-336v, Report by C. van Heijningen on trip to Canton, at 332-332v.

15. *P'ing-ting san-ni fang-lueh* (*TW*, No. 284), 27:211-212, 31:239, 35:272; Kai-fu Tsao, "The Rebellion of the Three Feudatories Against the Manchu Throne in China, 1673-1681: Its Setting and Significance" (PhD dissertation, Columbia University, 1965), pp. 127-136.

16. *Sinica Franciscana* (Quaracchi-Firenze and Rome, 1929-1965), III:170-185, Buenaventura Ibañez, O.F.M., Macao, to Father Provincial, January 1678, p. 182. Ibañez thought a lion and a lioness had been sent, but the lioness had died at sea; Verbiest, writing before the embassy reached Peking, also thought two lions had started the voyage and only one arrived; F. Verbiest, S.J., to King of Portugal, 7 September 1678, *Correspondance*, pp. 256-266, at 262. But the Goa sources make it clear that only one lion had been sent from there.

17. Goa, Monções 39-40:4, Viceroy to Prince-Regent, 4 December 1674; Goa 1264:16-16v, Administrators of Portuguese India to City of Macao, 7 May 1678, and 1264:30-31v, Viceroy to City of Macao, 4 May 1679.

18. José de Jesus Maria, pp. 84-85; *SL:KH*, 76:3-3b.

19. On March 5, Ibañez was writing as if the embassy had already left; *Sinica Franciscana*, III:186-196, B. Ibañez, O.F.M., Macao, to Father Provincial, 5 March 1678, p. 191.

20. José de Jesus Maria, p. 86.

21. List of loans by Macao officials and citizens, and expenses of the embassy of Bento Pereira de Faria, printed by C.R. Boxer and J.M. Braga as an appendix to their edition of F. Pimentel, S.J., *Breve Relação*, pp. xxviii-xlii. Bertuccioli, p. 223, Note 1, cites a manuscript summary of this document in the Biblioteca Nacional, Lisbon.

22. For Verbiest's letter before the embassy arrived, see Note 16 above; *SL:KH*, 76:3-3b, records its arrival and the gift of the lion; on the honors to Verbiest, see Huang Po-lu, *Cheng-chiao feng-pao* (Shanghai, 1894), 77v-79v.

23. Hummel, pp. 924-925, 929-930; Spence, *Emperor*, Chapter 5; Silas Wu, *Passage to Power*.

24. Magalhaens, *Nouvelle Relation*, pp. 313-314, 344; Bertuccioli, p. 225, Note 11.

25. *SL:KH*, 76:4.

26. *LPTL*, banquet charts, 1b, and 198:4-5b; see also the fascinating depiction of this banquet in Okada Gyokuzan, *Tōdo meishō zukai* (Osaka, 1805), 1:27.

27. *Ta-Ch'ing hui-tien*, K'ang-hsi, 76:7b.

28. Buglio's essay is discussed and translated in Bertuccioli; see also Fang Hao, *Chung-Hsi chiao-t'ung shih* (Taipei, 1954), III:226, IV:112.

29. Bertuccioli, pp. 224-225, especially Notes 10 and 11; three of these poems are translated in Appendix C of this book.

30. Lu Tz'u-yun, *Pa-hung i-shih*, 2:2-2b; Chi Yun, *Ju-shih wo-wen*, in his *Yueh-wei ts'ao-t'ang pi-chi*, in *Pi-chi hsiao-shuo ts'ung-shu* (1936), 118; Wu Ch'ang-yuan, *Ch'en-yuan shih-lueh* (1876), 16:27-27b; Lo-shu Fu, *Documentary Chronicle*, pp. 456-457.

31. Verbiest, translated in Appendix B. Chin-shan might refer to Jade Peak Pagoda (Yü-feng-t'a) at Jade Fountain Hill (Yü-ch'üan-shan), sup-

posedly a replica of that on Chin-shan Island in the Yangtze; Carroll B. Malone, *History of the Peking Summer Palaces under the Ch'ing Dynasty* (Urbana, 1934), pp. 20-21.

32. Ajuda/JA 49-V-17:509-518v, Letters and testimonies from City of Macao to the Jesuits, at 509-509v (letter from Pereira de Faria, 14 November 1678); *Sinica Franciscana,* III:479-482, A. de San Pascual, O.F.M., Chinan, to Father Provincial, 27 November 1678, p. 481. Copies of translations of the edict to the King of Portugal are to be found at Palha #2, Ajuda/JA 49-V-17:518-518v, and José de Jesus Maria, p. 95.

33. The granting of banquets and presents was ordered only on November 2; *SL:KH,* 77:7b. See Appendix D for the gifts.

34. A. de San Pascual, O.F.M., in the passage cited in Note 32 above, wrote that the lion died 40 days after it reached Peking; José de Jesus Maria, p. 103, says it died 15 days after the envoy left, which probably should be 15 days *before.*

35. Appendix B; José de Jesus Maria, p. 103; J.B. du Halde, S.J., *Déscription Géographique, Chronologique, Politique, et Physique de la Chine et de la Tartarie Chinoise* (The Hague, 1736), IV:298, which also refers to Manchu lore on medicinal virtues of parts of tigers.

36. Ajuda/JA 49-V-17:509-518v; see Note 32.

37. Petech, pp. 234-236.

38. José de Jesus Maria, pp. 90-104; only points drawn from other sources are noted separately below. See also Archivo Histórico Nacional, Madrid, Jesuitas, Leg. 270, num. 96, J. Tissanier, S.J., to A. Gomes, 29 January 1680. I am indebted to George B. Souza for sending me a copy of this document.

39. *P'ing-ting san-ni fang-lueh,* 36:280, 38:295, 296, 39:313, 40:319, 328.

40. VOC 1369:765v-779v, Dutch Ships in Macao Islands to Batavia, 20 November 1681, at 775-775v.

41. Hummel, p. 635; *Wu, Keng, Shang, K'ung ssu-wang ch'üan-chuan* (*TW,* No. 241), p. 26; Wills, "Maritime China," p. 232 and Note 64; ARSJ/JS 163:111-114, Summary of Jesuit accounts of the death of Shang Chih-hsin.

42. AHU, Macau, maço 1, Senate of Macao to Prince Regent, 19 December 1680.

43. Petech, p. 235; ARSJ/JS 125:164-199v, F. Couplet, S.J., to Cardinals of the Propaganda, "Breve relatione dello stato e qualita delle missione della Cina," p. 169; Ajuda/JA 49-V-17:a511-514v, B.P. de Faria to F. Verbiest, S.J., 20 January 1680 and 19 October 1680.

44. See the AHU letter cited in Note 42 above.

45. AHU, Macau, maço 1, Senate of Macao to Prince Regent, 5 December 1682; VOC 1377:601-619, Report at Batavia on voyage to Macao Is-

lands, 1 April 1682; VOC 1386:690v-701, Report at Batavia on voygage to Macao Islands, 31 March 1683; Goa 1264:49v-50, Antonio Paes de Sande to City of Macao, 10 May 1681. Li Shih-chen, *Fu-Chiang fu-Yueh cheng-lueh,* 7:16a-18b, confirms that Macao's land trade was legalized and subjected to taxation in 1680, with the understanding that the sea trade would be opened once the "sea rebels" had been pacified. Li was Governor of Kwangtung in the early 1680s.

FIVE: VINCENT PAATS, 1685-1687

1. I hope to give a full account of these developments in a study of changes in Sino-European trade in the 1680s. A fairly good short account of the Chinese changes of 1683-1685 is in my "Ch'ing Relations with the Dutch, 1662-1690" (PhD dissertation, Harvard University, 1967), pp. 512-520.

2. VOC 1415:1041-1043, Translation of imperial edict, dispatched on 21 January 1685.

3. VOC 1415:957-960v, Short report by Alexander van 's Gravenbroeck, 29 May 1685.

4. Wills, *Pepper,* p. 99.

5. *SL:KH,* 116:8a-b.

6. VOC 700:41, 214-216, 219-220, 230-231, 282-287, 288-289, 292-293, 304, 314, 317, 318, 326-328, Batavia resolutions 24 January-13 July 1685; VOC 912:474-528, Commission and instructions for V. Paats, letters to Emperor and Fukien officials, 20 July 1685; VOC 1408:9-927, Batavia to Seventeen, 11 December 1685, at 72v-85; VOC 1408:856-857v, V. Paats to Seventeen, 18 July 1685. See also J. Vixseboxse, *Een Hollandsch Gezantschap naar China in de Zeventiende Eeuw, 1685-1687* (Leiden, 1946), pp. 15-39, 115-121.

7. Abraham J. van der Aa, *Biographisch Woordenboek der Nederlanden* (Haarlem, 1852-1878), XV:17-20; P. Geyl, *The Netherlands in the Seventeenth Century, Part Two, 1648-1715* (New York and London, 1964), pp. 219, 250, 260; India Office Library and Records, London, Factory Records Java, Vol. 8, Papers pertaining to the English and Dutch Commissioners, 1684-1686; personal communication from Dr. B. Woelderink, Gemeente Archief, Rotterdam, 4 September 1970.

8. Vixseboxse, p. 20.

9. Wills, *Pepper,* p. 164.

10. For a detailed comparison of the Dutch and Chinese lists of the presents, see Vixseboxse, pp. 30-35.

11. On the first part of the stay in Foochow, see VOC 1415:965-1021, Foochow day-register, 25 August-15 October 1685; VOC 1429:823-834, Embassy to Batavia, 20 February 1686; Vixseboxse, pp. 40-60.

12. *HTSL,* 511:2-2b forbidding provincial officals to communicate privately with foreigners; *SL:KH,* 115:24b-25b on the Siamese precedent; *SL:KH,* 120:24b-25 and Vixseboxse, p. 53, on the exemption from tolls.

13. Vixseboxse, p. 58.

14. The most important records for the rest of the embassy are VOC 702:262, Batavia resolution 16 May 1687; VOC 1429:823-834, Embassy to Batavia, 20 February 1686; VOC 1431:557-563, Translation of Chinese record of communications between V. Paats and the Board of Ceremonies; VOC 1438:674-699v, Embassy to Batavia, 20 April 1687, and report by V. Paats, 24 February 1687; VOC 1438:702v-710, Translations of imperial edict to Governor-General Camphuys (including list of presents), 26 August 1686, Chinese record of communications between V. Paats and Board of Ceremonies (identical to VOC 1431:557-563), reports by Governor-General Wang Kuo-an and Board of Ceremonies and imperial decision on routes of embassies, 11 September 1686, document from Board of Ceremonies listing permitted and forbidden goods for future Dutch tribute, 22 August 1686, Board of Ceremonies record of receipt of Dutch presents and dispatch of imperial presents, imperial decision on interval of Dutch tribute and their request for a permanent lodge, n.d., Governor Chin Hung to Governor-General Camphuys, n.d.; VOC 1438:711-724, Report at Batavia by V. Paats and J. Leeuwenson, 6 May 1687, and appendixes.

15. VOC 1438:736-742v, Paats's instructions to Van 's Gravenbroeck, 15 November 1685, and report by Van 's Gravenbroeck, c. 10 December 1685.

16. VOC 1432:596-608v, Miscellaneous statements by Paats, Van 's Gravenbroeck et al; VOC 1431:609-617v, Statement by Van 's Gravenbroeck, n.d., before 22 December 1687 (the most informative part of these documents); VOC 1432:545-622v, Extracts from the resolutions of the Batavia Council and Statements by Paats, Van 's Gravenbroeck et al.; VOC 1432:632-643v, 659-663, VOC 1443:1042-1049v, Statements by members of the Batavia Council and resolutions of the Batavia Council; VOC 1438:699, Report by V. Paats, 24 February 1687; VOC 1458:1008-1010v, Statement by Van 's Gravenbroeck. (Some of these are also in VOC 702 and 703, Batavia resolutions for 1687 and 1688.) Points taken from these sources are noted separately below.

17. These day-registers did reach the Netherlands; they were quoted in Nicolaes Witsen, *Noord- en Oost-Tartaryen* (Amsterdam, 1692), pp. 39-40.

18. VOC 1431:599, 612-613v, Statements by Paats and Van 's Gravenbroeck.

19. VOC 1432:570, Statement by Van 's Gravenbroeck. The embassy's arrival and audience are recorded in *SL:KH,* 126:26a-b. See also Vixseboxse, pp. 61-81, and ARSJ/JS 104:305, 105 I:88v-89v, Dunin Szpot.

20. J. Bouvet, *Histoire de L'Empereur de la Chine* (The Hague, 1699; reprint, Tientsin, 1940), p. 128; Bosmans, "Verbiest . . Directeur," p. 419.

21. Tsiaulauja (Chao-lao-yeh) and variations in Dutch sources. Father Francis A. Rouleau, S.J., and Jonathan Spence pointed out this identification to me in personal communications. On his conversations with Paats, see VOC 1432:569-570, Statement by Van 's Gravenbroeck.

22. VOC 1438:691, Report by V. Paats, 24 February 1687.

23. BOC 1432:574, 577, Statements by Van 's Gravenbroeck; VOC 1432:598, Statement by Inspector-General Jeremias Coesart, 3 September 1687.

24. *SL:KH* 127:10-11; *K'ang-hsi yü-chih wen-chi* (Taipei, 1966), 4:10b-11; *P'ing-ting lo-ch'a fang-lueh,* in *Huang-ch'ao fan-shu yü-ti ts'ung-shu* (Shanghai, 1903), 24b-25.

25. VOC 1438:174, Batavia to Seventeen, 23 December 1687.

26. One Jesuit was sufficiently impressed by this source of influence at court to suggest to the General in Rome that, as long as the Dutch maintained relations with China, there ought to be some Flemish Fathers at Peking to translate for them; ARSJ /JS 162:277-280v, Francesco Brancato, S.J., Canton, 15 February 1669, at 279v.

27. Lo-shu Fu, *Documentary Chronicle,* pp. 469-470; Verbiest, *Correspondance,* pp. 528-531; Bosmans, "Verbiest . . Directeur," pp. 441-443; H. Bosmans, S.J., "Le problème des relations de Verbiest avec la Cour de Russie," *Annales de la Société d'Emulation pour l'étude de l'Histoire et des Antiquités de la Flandre,* 63:193-223, 64:98-101 (1913-1914), especially 63:208-210.

28. VOC 1440:2319v-2324, Extracts from F. Flettinger's daily notes, 12 February-30 August 1687, translated in Wills, "Some Dutch Sources on the Jesuit China Mission;" VOC 1431:176-176v, Batavia to Seventeen, 23 December 1687; VOC 1432:562, Statement by Van 's Gravenbroeck, 5 March 1688. For departure date, see ARSJ /JS 150:123-150, A. Thomas, S.J., "Annotationes . . . in Sina" at 123v-124.

29. VOC 1432:579v, 605v, Statements by Van 's Gravenbroeck and L. de Keyser.

30. VOC 1438:723v, Report by V. Paats and J. Leeuwenson, 6 May 1687; Vixseboxse, pp. 97-103; ARSJ/JS 150:149, A. Thomas, S.J., "Annotationes"; Arsip Nasional Republik Indonesia, Day-register of Batavia Castle for 1687, pp. 297-303, 317-322, dated 12 and 20 May 1687. Paats reported that the embassy had had imperial permission to trade at Hangchow on its way south but had been forbidden to do so by their accompanying officials; VOC 1438:678v, Report by V. Paats, 24 February 1687. I think it most unlikely that they had any such permission.

31. Fang Hao, *Chung-Hsi chiao-t'ung shih,* V:8-15; Verbiest, *Correspondance,* pp. 366-367; Bouvet, pp. 14, 83-84; Pfister, pp. 381-385.

32. Verbiest, *Correspondance*, pp. 488-495, 518-520.

33. VOC 1438:693, Report by V. Paats, 24 February 1687.

34. VOC 1438:740, Report by Van 's Gravenbroeck, c. 10 December 1685.

35. VOC 1440:2323v, Flettinger's daily notes, 11 August 1687.

36. VOC 1438:684-684v, Report by V. Paats, 24 February 1687; Vixseboxse, pp. 78-79.

37. VOC 1438:686v-687, Report by V. Paats, 24 February 1687. Paats here seems to echo Verbiest's view on the Emperor's unwillingness to reject a board recommendation; Bosmans, "Verbiest Directeur," p. 449.

38. *SL:KH*, 126:26a-b, 127:6b, 127:9b-10, 127:10-11.

39. Lo-shu Fu, *Documentary Chronicle*, pp. 469-470, Note 229; see also Yü Cheng-hsieh, *Kuei-ssu lei-kao* (1833; Shanghai reprint, 1957), 9:335.

40. *Ta-Ch'ing hui-tien*, K'ang-hsi, 72:14, 74:8b; *HTSL*, 502:16b, 502:20b, 503:6-7, 506:7b, 511:2b; *LPTL*, 178:2a-b, 178:3b; Liang T'ing-nan, *Yueh hai-kuan chih*, 22:12-13b; Liang T'ing-nan, *Yueh-tao kung-kuo shuo*, 3:14b-16b; *Ch'ing-ch'ao wen-hsien t'ung-k'ao*, 298:7473; *Ch'ing-ch'ao t'ung-tien* (Shanghai, 1936), 98:2741; *K'ang-hsi-ti yü-chih wen-chi*, 4:10b-11; Vixseboxse, pp. 73, 79-81, 122; Lo-shu Fu, *Documentary Chronicle*, pp. 85-86, 88; Liu Hsien-t'ing, *Kuang-yang tsa-chi*, reprinted in *Chung-kuo hsueh-shu ming-chu* (Taipei, n.d.), pp. 17-18; Wang Shih-chen, *Ch'ih-pei ou-t'an*, 1:59-60; Yü Wen-i, *Hsu-hsiu T'ai-wan fu-chih* (*TW*, No. 121), pp. 691-692.

41. Wang Shih-chen, *Ch'ih-pei ou-t'an*, 1:59; modified from Lo-shu Fu, *Documentary Chronicle*, p. 469; Chang Ju-lin and Yin Kuang-jen, 34b; also mistakenly recorded among the records of the Van Hoorn embassy in Liang T'ing-nan, *Yueh-tao kung-kuo shuo*, 3:12b. Wang Shih-chen in the same collection, 1:64, recorded a fulsome tribute memorial from Siam. Could it be that the great poet *wrote* these little shams?

SIX: THE SURVIVAL OF CH'ING ILLUSIONS

1. Wills, *Pepper*, pp. 204-212; See also the works cited in Chapter 1, Notes 3 and 4.

2. On seals, see the passage from Paats's report quoted at the end of Chapter 5. *SL:KH*, 76:3b, translated without comment the Western date of the "tribute memorial" brought by the Pereira de Faria embassy.

3. For a survey and citations on sources of eighteenth-century trade and diplomacy, see John L. Cranmer-Byng, and John E. Wills, Jr., "Trade and Diplomacy with Maritime Europe, 1644-c. 1800," *Cambridge History of China*, Volume 9, in press.

4. Liang T'ing-nan, *Yueh-tao kung-kuo shuo,* 1:1–2; see also the biography of Liang in Hummel, *Eminent Chinese,* pp. 503–505.

APPENDIX A

1. F. Pimentel, S.J., *Breve Relação,* ed., C.R. Boxer and J.M. Braga (Macao, 1942).

2. The Jesuit view of Nanking as a slough of immorality probably was derived in part from the views of Confucian moralists of the time. See also a reference to the "customs of Venus and Bacchus" there in *ARSJ/JS* 122:204–242, Antonio Gouvea, S.J., to Father General, Annual for 1644, dated 16 August 1645, at 208. On the term *sangley* for the Manila Chinese, see C.R. Boxer, ed. and tr., *South China in the Sixteenth Century* (London, 1953), p. 260. The text refers to Cheng Ch'eng-kung's attack on Nanking in 1659.

3. fortes e guerreros.

4. I-kuan was the name by which Cheng Chih-lung was known to Westerners.

5. Here and in the following paragraphs, *I-kuan* refers to Cheng Chih-lung's son Cheng Ch'eng-kung; twice the Boxer-Braga edition has "Levão" for Ajuda's "Icuão".

6. Eighteen thousand taels might be possible for the salt revenue of that area, but not 18 million.

7. Some officials also condemned this policy as ineffective and brutal, but it seems to me that it was effective in making the Cheng coastal bases untenable; see Wills, *Pepper,* pp. 16–17.

8. November of *1666;* see Chapter 3.

9. dispençassão.

10. Presumably the *shui-shih t'i-tu,* provincial commander of water forces.

11. But this must have included some servants or slaves of the other Portuguese; later in this text, Father Pimentel has a good deal to say about his own slave.

12. nenhum Embaxador entrou se não com titulo de tributario.

13. In this passage, *river* refers to both natural rivers and the Grand Canal.

14. See Chapter 2 for a brief description and citations on the flash-locks on the Grand Canal.

15. sirgua

16. e outras de ouro

17. que nos podiamos passar

18. amarras

19. *Apparently* this refers to the boat of the brother of the Fukien ruler, but the passage is not completely clear.

20. *K'an-ho* was an old Ming term for the travel and authentication documents of an embassy, especially well known in its application to embassies from Japan to the Ming court.

21. não se uza nesta gente

22. This long paragraph is not in the Ajuda text. The Palha text is as follows: Em todas as villas e cidades aonde chegou, o mandarim tem Ospedaria Real aonde recebem, e fazem os gastos a conta do Emperador, mais ou menos conforme a dignidade da pessoa, porem o Snr Embayxador nunca quis entrar em nenhum destas Ospedarias, mas ficar nas barcas, e na verdade tem tambem commodo q escezão a da terra, por q a barca tem no meyo duas camaras, e hua sala muy capazes com seus bofetes e cadeyras são pintadas e douradas com mil brincos e lavoras, todas rasgadas em genellas, q nos tempos asperos se fecham com portas de madeyra, nos tempos brandos, e suaves, se abrem as portas, e ficão as genellas cubertas com huns los que a modo arrezedo de volantes mas mais fortes e encorpados pelos quais o Mandarim esta vindo tudo q passa fora sem ser visto de ninguem. p[a] poppa e p[a] proa ficam outras camaras para os criados do mandarim sobree estas mais outro andar por sima de cazas mais bayxas, em humas se recolhem os marinheyros, e noutras as q tocão as Trombettas ao redor desta cazaria por toda a barca vay hum corredor de sinco palmas de largura por onde se serve e comonica toda a barca. Estas barcas são mayores ou menores conforme a capacidade dos Rios, mas as pequininas tem a mesma comedidade p[a] o Mandarim ainda q espaço mais comidade [copyist's error for pequeno?]. De Nancham metropoli de Kiansi athe a corte sam as barcas tamanhas que parecem Naos. Nos Rios não falta nada, não sendo necessario p[a] a vida, mas tambem p[a] o regalo. Bem se pode dizer com verdade que la da Cidade da China val por duas hua na agua, outra na terra, por q são tantos as barcas, e sam infinita a gente q nellas sirve q se não pode contar. Uns grandes, outros pequenos, de mil modos e feições per entro. Mas andam muitos arrependo [copyist's error for vendendo?] todo necessario p[a] o sustento outros por escizão o trabalho de pregam trazem certos sinais e bandeyrinhas q ensenando ja sabem o que vendem. Item costume achey nesta materia digno de grande reparo e he q q[do] algua fazendas poem a posta p[a] se vender poem lhe em sima hua crus de sorte q a cruz na China he sinal de Pregua q esta esposta a quem a quer comprar, uzo he este de santa crus capas de muyta concideração p[a] o q a conhece. [Pimentel's point here is that the heathen Chinese seem to have some anticipation of the Christian symbolism of the cross as an offer of salvation available to whoever will accept it.] Pertudo hum abundancia do peixe na quantidade e qualidade parece não pode ser mayor. Da lagoa q comessa em velem athe Gankin, q he hua cidade, q esta antes de chegar a Nankim, q são mais de quarenta legoas se pesca aquelle famoze peixe a q em Portugal chamão sothe [Uncertain reading; probably a sturgeon. The

1727 Portuguese embassy party noticed the same fish in the same area; see J.F. Judice Biker, *Collecção de Tratados e Concertos de Pazes que o Estado da India Portugueza Fez. . . . nas Partes da Asia e Africa Oriental* (Lisbon, 1881 et seq.), 6:120–121.], he aqui e de estranha grandeza, por alguns pezão mais de nove arrobas, nos mandamos comprar hua posta de ventreila q tinha mais de hum palmo ao compreido do peixe, e pezoze sessinta cates q são cem arrateis dos nossos, e preço he muyto barato, por que cada cate custou hum conderim e meyo q tem a ser na nossa moeda dous arrateis por hum vintem ainda menos; Com ser tam grandes não tem espinhas senão huns talos, a cartilages m̄ moles, por fora não he tudo cuberto de escamas, mas tem algumas muyto grandes q parecem laminas m^{to} fortes contudo parece carne de vaca por extreme gorda. Sendo pois as barcas tam comedas [comodas] e os Rios tam cheyos de todo o necessario p^{a} vida e p^{a} o regalo vam e vem os Mandarins de humas Provincias para outras com toda sua caza tam devagar, que mais parece se vem a recrear de que fazar [fazer] viagem por isso não sentem os trabalhos de caminho.

23. Here the Palha-only section ends and the Ajuda text resumes.

24. no *necessarios* in Palha text.

25. cazas

26. *Menos de oito dias* is in both texts, but must be a copyist's error for *mais de oite annos.*

27. ajoelhar tres vezes, e tornar se a pee, e cada vez q ajoelha se bayxa trez vezes a cabeça athe chegar a cham com a aba do chapeu.

28. que no mesmo pateo fica

29. *em pee,* not *sempre*

30. havendo se ja recolhido digo vindo meyo recolhendo o Snr Embaixador

31. Reys e Embaixadores

32. *lavar,* not *levar.* The end of the paragraph is impossible in both texts; perhaps *ahi o poem em albare.* "Albare" may be "albarelo," a large vase or jar. I am grateful to C.R. Boxer for help with this passage.

33. apontarey

34. *leitoens,* "suckling pigs," may be a copyist's error for *falcões.*

35. *lilibeos;* I cannot trace this word, but the context would suggest that it means "dwarfs."

36. These are, of course, the well-known North Chinese *k'ang.*

37. hum palmo, digo hum covado (Ajuda). The winter of 1670–1671 was exceptionally severe; see Jonathan D. Spence, *The Death of Woman Wang* (New York, 1978), p. 33.

38. Chao-ch'ing is in Kwangtung but was the seat of the Governor-General of Kwangtung and Kwangsi.

39. *officios* must be a copyist's error for *edificios.*

40. responderem a nenhum Rey
41. em sima (Ajuda)
42. reading *athe nas barbas* for *athe nas barcas.* The thought seems to be that women could show themselves and ride astride in this way because they were indistinguishable from men, because the men had no facial hair.
43. q˜ eram sete
44. tam naturaes. Andre Coelho Vieyra's appointment as second-in-command of the embassy is recorded in Goa 1210:27-27v, dated 30 January 1669. Goa 1210:40 seems to say that he was left in Canton when the embassy went to Peking; this probably is a clumsy fake by Bento Pereira de Faria, who was compiling this letter-book.
45. aos outros?
46. *sucesso* read as copyist's error for *successão*
47. p^{a} a sua barca
48. capaz de pena e sentimento
49. sette e meyo
50. vos as aseytais
51. porem elle nenhum cazo fez das ordens do Cap^{m} General, entrou em Macau de mesmo modo q˜ sempre; veyo diante com a bandeira
52. o pouco respeito que tem a tantos
53. se vinha tratar algum negocio sobre Macao
54. This would imply that, on the journey north, Saldanha received gifts from various local officials but did not visit them.
55. *opia* probably is a copyist's error for *optima.*
56. The Peking Jesuits did translate a letter from the Dutch in 1668-1669, but I cannot see what the "equivocation" can have been or that any change in wording would have made much difference. See Wills, *Pepper,* p. 139, note 35.
57. Perhaps some toll barrier, *kuan,* in the T'ung-chou area.
58. não só de palavra
59. em esta Fee foi recebida a embaixada (Ajuda)
60. Probably Mingju and Songgotu, respectively.
61. Chang-chou, here and frequently used for southern coastal Fukien. The events mentioned here are described in Wills, *Pepper,* Chapter 2.
62. logo Bento Pereyra se queyxou dos Padres e os apertou q elles lhe disserão
63. que os Padres me derão
64. *Son of the sea* would be *yang-tzu,* but not with the usual characters for the Yangtze River.
65. hum grande serpente q vivia
66. todos são seus criados q fazem o q˜ lhe elle manda, assim como na China não ha mais q˜ hum Emperador, todos mais são seus criados, e vassalos.
67. "Brigade general" is the standard translation.

68. saltavão do altar
69. T'ung Kuo-ch'i; see Chapter 3, Note 85.
70. Copyist's error for *contra Reyes alheos?*
71. Philippians 1:15 and 1:18
72. Copyist's error for *execrado?*
73. The royal standard bore the arms of the great Portuguese knightly order, of which Henry the Navigator had been a Grand Master.
74. amigo e fautor
75. a dignidade, digo a idade
76. A letter by François de Rougemont, S.J., dated 11 March 1671 is the earliest mention I have found of permission to return to churches in the provinces; ARSJ/JS 162:324–325v.

APPENDIX B

1. H. Bosmans, "Verbiest . . . Russie," pp. 202.
2. I have found no trace of this document; it probably was sent on to the King of Portugal, and was lost as a result of the great Lisbon earthquake and the general lack of archival organization in the eighteenth century. Carlos Sommervogel, *Bibliothèque de la Compagnie de Jésus* (Brussels, 1890–1932), Vol. 3, Column 1835. Verbiest, *Correspondance*, pp. 334–335.
3. This translation is based in part on the phrasing of T.I. Dunin Szpot, S.J., ARSJ/JS 104:306: "ut ipsis e sanguine suo Regio Principibus Regulisque (quibus tamen nostros Patres spectatores interesse voluit) vix semel copiam eius videndi fecerit."
4. Dunin Szpot, ARSJ/JS 104:306v: "ut tandem et ij liberam navigandi quaquaversum ad maritimarum urbium portus potestatem, et in Sinam interiorem promovenda sua commercia obtinuerint."
5. Hummel, pp. 577–578.

APPENDIX C

1. These three poems are recorded in the 1751 edition of Chang Ju-lin and Yin Kuang-jen, *Ao-men chi-lueh*, 2.17a–b. They are omitted from the *Chao-tai ts'ung-shu* edition of this work. The translations are my own, done with the advice and correction of Joseph Chen and Dominic C.N. Cheung.

APPENDIX D

1. VOC 1438:699, Report by V. Paats, 24 February 1687.

2. *HTSL* 506:7.

3. *Ta-Ching hui-tien,* K'ang-hsi, 74:7b–11b, 77:20b–22; *LPTL,* 178:3b–4b, 180:4–5, 200:8–10; *HTSL,* 506:3–6b, 520:6a–b, 520:8a–b.

4. Petech, p. 232; ARSJ/JS 124:81–86v, "Novas da China, Anno 1669," at 85v; VOC 1265:1074v, 1085v–1086, 1089v, Embassy day-register, 22 June, 17 July, and 29 July 1667, printed in Dapper, pp. 353, 356–357, 367–368; for a partial list of the presents to the King of Portugal in 1670 and a full list for 1678, see the translations of edicts cited in Chapter 3, Note 118, and Chapter 4, Note 32.

APPENDIX E

1. *Ta-Ch'ing hui-tien,* K'ang-hsi, 72:19a–b; *LPTL,* 178:2–3, 180:2; *HTSL,* 503:4b–7; Pimentel (Boxer-Braga edition), v–vii; VOC 912:501–503, Batavia to Emperor of China, 20 July 1685; Vixseboxse, pp. 30–35, 119–121. A Chinese list of the Van Hoorn presents is preserved in Portuguese translation in Ajuda/JA, 49:V:15, fol. 339–341. The reception of the horses and oxen in Peking is discussed in Chapter 2; the glass lantern and the globe and celestial sphere, possibly a single globe with astronomical additions (*hemel- en aert-kloot*) are mentioned in VOC 1258:1472, Embassy day-register 14 October 1666, printed in Dapper, p. 262.

2. *Ta-Ch'ing hui-tien,* K'ang-hsi, 72:13a–b; *LPTL,* 178:1b–2.

Bibliography

NOTE ON SOURCES

This book is based on substantial quantities of manuscript records in six archives and scraps of information from three others. Contemporary printed works, especially those by Jesuits, also provide information not duplicated in any known archival material. The challenges of finding these rare old books and the passages in them relevant to a particular topic frequently are as great as those of finding and using archival materials. For example, I have consulted the *Lettres des Pays Estrangers* of Philippe Chahu, S.J., in microfilm from the British Library; I do not believe there is a copy in the United States.

Jesuit manuscript sources have been found primarily in the papers of the Japan-China Province in the Archivum Romanum Societatis Jesu and in the "Jesuitas na Ásia" series in the Biblioteca da Ajuda, Lisbon. The latter is an extremely important collection of eighteenth-century copies, made at the time of the dissolution of the Society of Jesus in the Portuguese empire, that should be consulted on any topic in Far Eastern mission history at least down to the 1680s.

Two collections of Portuguese government documents have been consulted: the Arquivo Histórico Ultramarino, Lisbon, especially the chronologically arranged "bundles" (maços) on Macao; and the Historical Archives in Goa. For these embassies, this material, while valuable, is far less full than that from Jesuit sources.

For explanation of the Dutch sources, see my *Pepper*, pp. xiii–xv. For more details on the holdings of all these archives, see Wills, "Early Sino-European Relations: Problems, Opportunities, and Archives," *Ch'ing-shih wen-t'i*, 3.2:50–76 (December 1974), and Wills, "Advances and Archives in Early Sino-Western Relations: An Update," *Ch'ing-shih wen-t'i*, 4.10:87–110 (December 1983). For abbreviations and numbering systems used in this book see "Abbreviations" preceding the Notes.

ARCHIVAL SOURCES

England. India Office Library and Records, London
India. Historical Archives, Panjim, Goa
Indonesia. Arsip Nasional Republik Indonesia, Jakarta

Italy. Archivum Romanum Societatis Jesu, Rome
Netherlands. Algemeen Rijksarchief, The Hague
Portugal. Arquivo Histórico Ultramarino, Lisbon
Portugal. Biblioteca da Ajuda, Lisbon
Spain. Archivo Histórico Nacional, Madrid
United States. Houghton Library, Harvard University, Cambridge

PRINTED SOURCES

Aa, Abraham J. van der. *Biographisch Woordenboek der Nederlanden.* 21 vols. Haarlem, 1852–1878.

Bernet Kempers, A.J., ed. *Journaal van Dircq van Adrichem's Hofreis naar den Groot-Mogol Aurangzeb, 1662.* Works of the Linschoten Vereeniging, XL. The Hague, Nijhoff, 1941.

Bertuccioli, Giuliano. "A Lion in Peking: Ludovico Buglio and the Embassy to China of Bento Pereira de Faria in 1678," *East and West* (Rome), New Series, 26.1–2:223–240 (1976).

Bosmans, H., S.J. "Ferdinand Verbiest, Directeur de l'Observatoire de Peking (1623–1688)," *Revue des Questions Scientifiques* (Brussels), 3d series, 21:195–273; 21:375–464 (1912).

——. "Le problème des relations de Verbiest avec la Cour de Russie," *Annales de la Société d'Emulation pour l'étude de l'Histoire et des Antiquités de la Flandre,* 63:193–223; 64:98–101 (1913–1914).

——. *Lettres Inédites de Francois de Rougemont.* Louvain, Bureaux des Analectes, 1913.

Bouvet, J. *Histoire de l'Empereur de la Chine.* The Hague, M. Uytwerf, 1699, reprinted in Tientsin, 1940.

Boxer, C.R. "Portuguese Military Expeditions in Aid of the Mings Against the Manchus, 1621-1647," *T'ien-hsia Monthly,* 7.1:24–50 (August 1938).

——. "The Rise and Fall of Nicholas Iquan," *Tien-hsia Monthly,* 11.5:401–439 (April-May 1939).

——. *Macau na Época da Restauração (Macau Three Hundred Years Ago).* Macao, Imprensa Nacional, 1942.

——. *Portuguese Society in the Tropics: The Municipal Councils of Goa, Macao, Bahia, and Luanda, 1510–1800.* Madison and Milwaukee, University of Wisconsin Press, 1965.

——. *Francisco Vieira de Figueiredo: A Portuguese Merchant-Adventurer in South East Asia, 1624–1667.* Verhandelingen van het Koninklijk Instituut voor Taal-, Land-, en Volkenkunde, No. 52. The Hague, Nijhoff, 1967.

——. *Fidalgos in the Far East.* 1948; reprinted, London, Oxford University Press, 1968.

Boxer, C.R., ed. and tr. *South China in the Sixteenth Century.* Hakluyt Society, Series II, No. 106. London, 1953.

Brazão, Eduardo. "The embassy of Manuel de Saldanha to China in 1667–1670 (Notes on Sino-Portuguese Diplomatic Relations)," *Boletim do Instituto Português de Hongkong,* no. 1. July 1948, pp. 139–162.

——. *Subsídios para a História das Relaçes Diplomáticas de Portugal com a China: A Embaixada de Manuel de Saldanha, 1667–1670.* Macao, Imprensa Nacional, 1948.

——. *Apontamentos para a História das Relações Diplomáticas de Portugal com a China, 1516–1753.* Lisbon, Agência Geral de Colónias, 1949.

Cardim, Antonio Francisco, S.J. *Batalhas da Companhia de Jesus na sua Gloriosa Provincia de Japão.* Lisbon, Imprensa Nacional, 1894.

Chahu, Philippe, S.J. *Lettres des Pays Estrangers.* Paris, Denis Bechet, 1668.

Chang Ju-lin 張汝霖 and Yin Kuang-jen 印光任. *Ao-men chi-lueh* 澳門紀略 (A summary of Macao). 1751; also reprinted in *Chao-tai ts'ung-shu* 昭代叢書 (Chao-tai collected works). n.d.

Chang Wei-hua 張維華. *Ming-shih Fo-lang-chi, Lü-sung, Ho-lan, I-ta-li-ya ssu-chuan chu-shih* 明史佛郎機呂宋荷蘭意大里亞四傳注釋 (A commentary on the four chapters on Portugal, Spain, Holland, and Italy in the *History of the Ming Dynasty*). Yenching Journal of Chinese Studies Monograph Series No. 7. Peiping, 1934.

Ch'en Wei-sung 陳維崧. *Hu-hai-lou shih-chi* 湖海樓詩集 (Collected poems from the Lake and Sea Pavilion). 1689.

Cheng-shih shih-liao san-pien 鄭氏史料三編. (Historical sources on the Cheng family, 3rd ser.). 2 vols. *TW,* No. 175.

Chi Yun 紀昀. *Ju-shih wo-wen* 如是我聞 (Thus I have heard). In his *Yueh-wei ts'ao-t'ang pi-chi* 閱微草堂筆記 (Notes from the Grass Hut for Examining Subtleties), in *Pi-chi hsiao-shuo ts'ung-shu* 筆記小說叢書 (Collection of jottings and stories). n.p., 1936.

Chinese Academy of Architecture. *Ancient Chinese Architecture.* Peking and Hong Kong, 1982.

Ch'ing-chao t'ung-chih 清朝通志 (Comprehensive essays for the Ch'ing dynasty). Shih-t'ung edition, Shanghai, 1936.

Ch'ing-ch'ao t'ung-tien 清朝通典 (Comprehensive institutions for the Ch'ing dynasty). Shih-t'ung edition, Shanghai, 1936.

Ch'ing-ch'ao wen-hsien t'ung-k'ao 清朝文獻通考 (Encyclopedia of historical institutions for the Ch'ing Dynasty). 4 vols. Shih-t'ung ed., Shanghai, 1936.

Ch'ing ch'i-hsien lei-cheng hsuan-pien 清耆獻類徵選編 (Selections from the Ch'ing *Classified Biographies of Worthies*). *TW,* No. 230.

Chiu T'ang-shu 舊唐書. (Old T'ang history). Po-na-pen ed.

Chou Ching-lien 周景濂. *Chung-P'u wai-chiao shih* 中葡外交史 (History of Sino-Portuguese relations). Shanghai, 1936.

Coolhaas, W. Ph., ed. *Generale Missiven van Gouverneurs-Generaal en Raden aan Heren XVII der Verenigde Oostindische Compagnie.* 5 vols. The Hague, Nijhoff, 1960–1975.

Cordier, Henri. *Bibliotheca Sinica.* 5 vols. Paris, Guilmoto, 1904–1908 and Paris, P. Geuthner, 1922–1924.

Cranmer-Byng, John L., and John E. Wills, Jr. "Trade and Diplomacy with Maritime Europe, 1644–c. 1800," in *Cambridge History of China,* Vol. 9, in press.

Creel, Herrlee G. *The Origins of Statecraft in China. Volume One: The Western Chou Empire.* Chicago and London, University of Chicago Press, 1970.

Crossley, Pamela. "The Tong in Two Worlds: Cultural Identities in Liaodong and Nurgan during the 13th–17th Centuries," *Ch'ing-shih wen-t'i,* 4.9:21–46 (June 1983).

Dagh-Register gehouden in 't Casteel Batavia, 1628–1682. 31 vols. Batavia, 1887–1931.

Dam, Pieter van. *Beschryvinge van de Oostindische Compagnie,* 7 vols. Ed. F.W. Stapel and C.W.Th. Baron van Asperen en Dubbeldam. The Hague, Nijhoff, 1927–1954.

Dapper, Olfert. *Gedenkwaerdig Bedryf der Nederlandsche Oost-Indische Maetschappye op de Kuste en in het Keizerrijk van Taising of Sina.* Amsterdam, Jacob van Meurs, 1671. English translation by John Ogilby, *Atlas Chinensis . . . . by Arnoldus Montanus,* London, 1671. See Chapter 2, Note 26 for fuller discussion.

Dubs, Homer H. *See* Pan Ku.

Elias, Johan E. *De Vroedschap van Amsterdam, 1578–1695.* 2 vols. Haarlem, Loosjes, 1903–1905.

Fairbank, John K. *Trade and Diplomacy on the China Coast.* 2 vols. Cambridge, Harvard University Press, 1953.

——. ed. *The Chinese World Order.* Cambridge, Harvard University Press, 1968.

—— and S.Y. Teng. "On the Transmission of Ch'ing Documents," *Harvard Journal of Asiatic Studies,* 4:12–46 (1939). Reprinted in Fairbank and Teng, *Ch'ing Administration: Three Studies,* Harvard-Yenching Institute Studies, 19. Cambridge, 1960.

——. "On the Ch'ing Tribute System," *Harvard Journal of Asiatic Studies* 6: 135–246 (1941). Reprinted in Fairbank and Teng, *Ch'ing Administration: Three Studies,* Harvard-Yenching Institute Studies, Cambridge, 1960.

Fang Hao 方豪. *Chung-Hsi chiao-t'ung shih* 中西交通史 (History of Sino-Western relations). 5 vols. Taipei, 1954.

Feenstra Kuiper, J. *Japan en de Buitenwereld in de Achttiende Eeuw.* The Hague, Nijhoff, 1921.

Fitzgerald, C.P. *Son of Heaven: A Biography of Li Shih-min, Founder of the T'ang Dynasty.* Cambridge, Cambridge University Press, 1933.

Fu Lo-shu, "The Two Portuguese Embassies to China during the K'ang-hsi Period," *T'oung Pao*, 43:75–94 (1955).

——. *A Documentary Chronicle of Sino-Western Relations (1644–1820).* 2 vols. Tucson, University of Arizona Press, 1966.

Geyl, P. *The Netherlands in the Seventeenth Century, Part Two, 1648–1715.* New York and London, Barnes and Noble and Ernest Benn, 1964.

Goodrich, L. Carrington, and Chaoying Fang. *Dictionary of Ming Biography, 1368–1644.* 2 vols. New York and London, Columbia University Press, 1976.

Greslon, Adrien, S.J. *Histoire de la Chine sous la Domination des Tartares.* Paris, Henault, 1671.

Grimm, Tilemann. *Erziehung und Politik im konfuzianischen China der Ming-Zeit, 1368–1644.* Hamburg, Harrassowitz, 1960.

Hann, F. de. *Priangan.* Batavia, 1910–1912.

Halde, J.B. du, S.J. *Déscription Géographique, Chronologique, Politique, et Physique de la Chine et de la Tartarie Chinoise.* 4 vols. The Hague, H. Scheurleer, 1736.

Ho Ch'ang-ling 賀長齡, ed. *Huang-ch'ao ching-shih wen-pien* 皇朝經世文編 (Collected essays on statecraft from the reigning dynasty). 3 vols. 1826; Taipei reprint, 1963.

Ho Yun-yi. "Ritual Aspects of the Founding of the Ming Dyansty, 1368–1398," *Bulletin of the Society for the Study of Chinese Religions,* 7:58–70 (Fall 1979).

Hoorn, Pieter van. *Eenige Voorname Eygenschappen van de Ware Deugdt, Wysheydt, en Volmaecktheydt, Getrocken uyt den Chineschen Confucius.* Batavia, Joannes van den Eede, 1675.

Hou Han-shu 後漢書. (History of Later Han). Po-na-pen edition.

Hsieh Kuo-chen 謝國楨. "Ch'ing-ch'u tung-nan yen-hai ch'ien-chieh k'ao" 清初東南沿海遷界考 (A study of the evacuation of the southeast coast in the early Ch'ing), *Kuo-hsueh chi-k'an* 國學季刊 (Sinological Quarterly), 2.4:797–826 (December 1930); English translation, *Chinese Social and Political Science Review,* 15:559–596 (1931).

Hsin Tang-shu 新唐書. (New history of the T'ang). Po-na-pen edition.

Hsu Tzu 徐鼒. *Hsiao-t'ien chi-nien* 小腆紀年 (Annals of a dynastic remnant). 5 vols. *TW,* No. 134.

Huang Po-lu 黃伯祿, *Cheng-chiao feng-pao* 政教奉褒 (Imperial favors received by the true faith). Shanghai, 1894.
Hummel, Arthur W., ed. *Eminent Chinese of the Ch'ing Period.* Washington, U.S. Government Printing Office, 1943–1944.

The I-li, or Book of Etiquette and Ceremonial. Tr. John C. Steele, London, Probsthain, 1917.

Jochim, Christian. "The Imperial Audience Ceremonies of the Ch'ing Dynasty," *Bulletin of the Society for the Study of Chinese Religions,* 7:88–103 (Fall 1979).
José de Jesus Maria, O.F.M. *Azia Sínica e Japónica.* Ed. C.R. Boxer. 2 vols. Macao, Escola Tipografica do Oratorio de S.J. Bosco (Salesianos), 1941, 1950.
Judice Biker, J.F. *Collecção de Tratados e Concertos de Pazes que o Estado da India Portugueza Fez com os Reis e Senhores com quem Teve Relações na Partes da Ásia e África Oriental.* 14 vols. Lisbon, Imprensa Nacional, 1881–1887.

K'ang-hsi-ti yü-chih wen-chi 康熙帝御製文集 (Collected writings of the K'ang-hsi Emperor). 4 vols. In *Chung-kuo shih-hsueh ts'ung-shu* 中國史學叢書 (Chinese historiography collection). Taipei, 1966.
Kessler, Lawrence D. *K'ang-hsi and the Consolidation of Ch'ing Rule, 1661–1684.* Chicago and London, University of Chicago Press, 1976.
Kuang-tung t'ung-chih 廣東通志 (Gazetteer of Kwangtung). Edition of Yung-cheng period.

Lao Chih-pien 勞之辨. *Ching-kuan-t'ang shih-chi* 靜觀堂詩集 (Collected poems from the Hall of Calm Contemplation). n.d.
Legge, James, tr. *The Chinese Classics.* 5 vols. Reprint, Hong Kong, Hong Kong University Press, 1960.
Li-chi 禮記 (Record of ceremonies). Tr. James Legge. Editorial matter by Ch'u Chai and Winberg Chai. New Hyde Park, University Books, 1967.
Li-pu tse-li 禮部則例 (Regulations and precedents of the Board of Ceremonies). 1841; Taipei reprint, 1966.
Li Shih-chen 李士禎. *Fu-Chiang fu-Yueh cheng-lueh* 撫江撫粵政略 (Records of governing in Chiang-nan and Kwangtung). n.d. Copy in Tōyō Bunko, Tokyo.
Li Yuan-tu 李元度. *Ch'ing hsien-cheng shih-lueh hsuan* 清先正事略選 (Selections from the Ch'ing *Notices of the Deeds of Past Worthies*). 2 vols. *TW,* No. 194.
Liang T'ing-nan 梁廷枏. *Yueh hai-kuan chih* 粵海關志 (Gazetteer of the Kwangtung Maritime Customs). c. 1839.

——. *Yueh-tao kung-kuo shuo* 粵道貢國說 (Account of countries that pay tribute via Kwangtung). In *Chung-kuo wen-shih ts'ung-shu* 中國文史叢書 (Collection on Chinese history and literature), No. 58, Taipei, 1968.

Lin, T.C. "Manchuria in the Ming Empire," *Nankai Social and Economic Quarterly*, 8.1:1–43 (1935).

Liu Hsien-t'ing 劉獻廷. *Kuang-yang tsa-chi* 廣陽雜記 (Kuang-yang miscellaneous notes). Reprinted in *Chung-kuo hsueh-shu ming-chu* 中國學術名著 (Famous works of Chinese scholarship). Taipei, n.d.

Liu, James T.C. *Ou-yang Hsiu: An Eleventh-Century Neo-Confucianist.* Stanford, Stanford University Press, 1967.

Ljungstedt, Sir Andrew. *An Historical Sketch of the Portuguese Settlements In China.* Boston, James Munroe, 1836.

Lu Tz'u-yun 陸次雲. *Pa-hung i-shih* 八紘譯史 (Translations and historical notes from the whole world). In *Lung-wei mi-shu* (Lung-wei confidential records), ts'e 65–66.

Mackerras, Colin, tr. and ed., *The Uighur Empire According to the T'ang Dynastic Histories: A Study in Sino-Uighur Relations, 744–840.* Columbia, University of South Carolina Press, 1972.

Magalhaens, Gabriel de, S.J. *Nouvelle Relation de la Chine.* Paris, Claude Barbin, 1688.

Maitra, K.M., tr. and ed. *A Persian Embassy to China: Being an Extract from "Zubdatu't-tawarikh" of Hafiz Abru.* Lahore, 1934; reprint New York, Paragon, 1970.

Malone, Carroll B. *History of the Peking Summer Palaces Under the Ch'ing Dynasty.* Illinois Studies in the Social Sciences, Vol. 19, No. 1–2. Urbana, 1934.

Marques Pereira, J.F. "Uma resurreição histórica (Páginas inéditas dum visitador dos jesuitas, 1665–1671), *Ta-Ssi-Yang-Kuo* (Macao), 1:31–41, 1:113–119, 1:181–188, 1:305–310, 2:693–702, 2:747–763 (1899, 1901).

Menezes, Luis de, Conde de Ericeira, *História de Portugal Restaurado.* 4 vols. Original ed. Lisbon, 1710; new ed., annotated by Antonio Alvaro Doria, Porto, Livraria Civilização, 1945.

Meng Sen 孟森. *Ming-tai shih* 明代史 (History of the Ming period). Taipei, 1957.

Ming-Ch'ing shih-liao 明清史料 (Ming and Ch'ing historical materials). Peking, Shanghai, and Taipei, 1930 et seq.

Ming hui-tien 明會典 (Collected statutes of the Ming). 1587; reprint, Kuo-hsueh chi-pen ts'ung-shu (Basic Sinological Series), Vol. 78–83, Taipei, 1968.

Ming hui-yao 明會要 (Collected essentials of the Ming), ed. Lung Wen-pin 龍文彬 Reprint, Peking, 1956.

Ming-shih 明史 (History of the Ming). Po-na-pen ed.

Montalto de Jesus, C. A. *Historic Macao.* Hong Kong, Kelly and Walsh, 1902.

Montanus, Arnoldus. *See* Dapper.

Moses, Larry W. "T'ang Tribute Relations with the Inner Asian Barbarian." In John C. Perry and Bardwell L. Smith, *Essays on T'ang Society: The Interplay of Social, Political and Economic Forces.* The Hague, Nihjoff, 1976.

Navarrete, Domingo Fernandez, O.P. *Travels and Controversies of Friar Domingo Navarrete, 1618–1686.* Ed. J.S. Cummins. 2 vols. Hakluyt Society, Series II, Vols. 118–119. London, 1962.

Needham, Joseph. *Science and Civilization in China.* Cambridge, Cambridge University Press, 1954 et seq.

Nieuhoff, Johan. *Het Gezantschap der Neerlandsche Oost-Indische Compagnie aan den Grooten Tartarischen Cham, Den Tegenwoordigen Keizer van China.* Amsterdam, Jacob van Meurs, 1665.

Ogilby, John. *See* Dapper.

Okada Gyokuzan 岡田玉山. *Tōdo meishō zukai* 唐土名勝圖會 (Collection of pictures of famous places in China). Osaka, 1805.

Oxnam, Robert B. *Ruling from Horseback: Manchu Politics in the Oboi Regency, 1661–1669.* Chicago and London, University of Chicago Press, 1975.

Pan Ku 班固. *Ch'ien Han-shu* 前漢書 (History of the Former Han). Po-na-pen edition. Also cited from Homer H. Dubs, tr., *History of the Former Han Dynasty.* 3 vols. Baltimore, Waverly Press, 1938, 1944, 1955.

Pelliot, Paul. "L'ambassade de Manoel de Saldanha a Pekin," *T'oung Pao,* 27:421–424 (1930).

Petech, Luciano. "Some Remarks on the Portuguese Embassies to China in the K'ang-hsi Period." *T'oung Pao,* 44:227–241 (1956).

Pfister, L., S.J. *Notices Biographiques et Bibliographiques sur les Jésuites de l'Ancienne Mission de Chine (1552–1773).* 2 vols. Variétés Sinologiques, Nos. 59, 60. Shanghai, 1932, 1934.

Pimentel, Francisco, S.J. *Breve Relação da Jornada que Fez à Corte de Pekim o Senhor Manoel de Saldanha, Embaixador Extraordinário del Rey de Portugal ao Emperador da China, e Tartaria (1667–1670).* Ed. C.R. Boxer and J.M. Braga. Macao, Imprensa Nacional, 1942.

P'ing-ting san-ni fang-lueh 平定三逆方略 (Summary of the pacification of the rebellion of the three feudatories). 3 vols. *TW,* No. 284.

P'ing-ting lo-ch'a fang-lueh 平定羅刹方略 (Summary of the pacification of the Russians). In *Huang-ch'ao fan-shu yü-ti ts'ung-shu* 皇朝

藩屬輿地叢書 (Collection of works on the geography of the border regions of the reigning dynasty). Shanghai, 1903.

Pires de Lima, Durval R. *A Embaixada de Manuel de Saldanha ao Imperador K'hang hi em 1667-1670 (Subsídios para a História de Macau).* Lisbon, Tipografia e Papelaria Carmona, 1930.

Pissurlencar, Panduronga S.S., ed. *Assentos do Conselho do Estado da India.* 5 vols. Bastorá, Tipografia Rangel, 1953–1958.

Rhede van der Kloot, M.A. van. *De Gouverneurs-Generaal en Commissarissen Generaal van Nederlandsch-Indië.* The Hague, Stockum, 1891.

Rossabi, Morris. "Ming China and Turfan, 1406-1517," *Central Asiatic Journal,* 16.3:206-225 (1972).

——. *China and Inner Asia: From 1368 to the Present Day.* London, Thames and Hudson, 1975.

——. *The Jurchens in the Yuan and Ming.* Cornell University East Asian Papers, No. 27. Ithaca, Cornell China-Japan Program, 1982.

——. ed. *China Among Equals: The Middle Kingdom and Its Neighbors, 10th–14th Centuries.* Berkeley, University of California Press, 1983.

Rougemont, François de, S.J. *Historia Tartaro-Sinica Nova.* Louvain, Hullegaerde, 1673.

Schwarz-Schilling, Christian. *Der Friede von Shan-yüan (1005 n. Chr.): Ein Beitrag zur Geschichte der Chinesischen Diplomatie. Asiatische Forschungen,* Band 1. Wiesbaden, Otto Harrassowitz, 1959.

Sebes, Joseph, S.J. *The Jesuits and the Sino-Russian Treaty of Nerchinsk (1689): The Diary of Thomas Pereira, S.J.* Bibliotheca Instituti Historici S.I., Vol. 18. Rome, Institutum Historicum S.I., 1961.

Serruys, Henry, C.I.C.M. *Sino-Jurčed Relations during the Yung-lo Period (1403-1424),* Göttinger Asiatische Forschungen, 4. Wiesbaden, 1955.

——. *Sino-Mongol Relations During the Ming, II, The Tribute System and Diplomatic Missions, Mélanges Chinois et Bouddhiques,* 14. Brussels, 1967.

Shu-ching 書經. Cited from James Legge, tr. *The Chinese Classics,* Vol. 3. Reprint, Hong Kong, Hong Kong University Press, 1960.

Silva Rego, Antonio da. "Macau entre duas crises (1640–1688)," *Anais da Academia Portuguesa de História,* 24.2:307-324 (1977).

Sinica Franciscana. 7 vols. Quaracchi-Firenze and Rome, 1929–1965.

Sirén, Osvald. *The Imperial Palaces of Peking.* Paris and Brussels, Van Oest, 1926.

Sommervogel, Carlos. *Bibliothèque de la Compagnie de Jésus.* 12 vols. Brussels, O. Schepens, 1890–1932.

Spence, Jonathan D. *Emperor of China: Self-Portrait of K'ang-hsi.* New York, Knopf, 1974.

——. "Ch'ing", in K. C. Chang, ed., *Food in Chinese Culture: Anthropological and Historical Perspectives.* New Haven and London, Yale University Press, 1977, pp. 259–294.

——. *The Death of Woman Wang.* New York, Viking, 1978.

Ssu-ma Ch'ien. *Records of the Grand Historian of China.* Tr. Burton Watson. 2 vols. New York, Columbia University Press, 1961.

Ssu-ma Kuang 司馬光. *Tzu-chih t'ung-chien* 資治通鑑 (A comprehensive mirror for the aid of government). 4 vols. Hong Kong, Chung-hua, 1956.

Sung-shih 宋史 (History of the Sung). Po-na-pen ed.

Ta-Ch'ing hui-tien 大清會典 (Collected statutes of Great Ch'ing). K'ang-hsi edition, 1690; Kuang-hsu ed., 1899, Taiwan reprint, 1963. [Intervening editions checked from Yung-cheng, Ch'ien-lung and Chia-ch'ing periods.]

Ta-Ch'ing hui-tien shih-li 大清會典事例 (Cases and precedents for the collected statutes of Great Ch'ing). 1899; Taiwan reprint, 1963. [Precedent material was incorporated in the 1690 *Hui-tien.* Earlier separate compilations of precedents are *Ta-Ch'ing hui-tien tse-li* 則例 of the Yung-cheng and Ch'ien-lung periods and *Shih-li* of Chia-ch'ing.]

Ta-Ch'ing hui-tien t'u 大清會典圖 (Charts for the collected statutes of Great Ch'ing). 1899; Taiwan reprint, 1963.

Ta-Ch'ing li-ch'ao shih-lu 大清歷朝實錄 (Veritable records of successive reigns of Great Ch'ing). Tokyo, 1937–1938; Taiwan reprint, 1963.

Ta-Ch'ing t'ung-li 大清通禮 (Comprehensive ceremonies of Great Ch'ing). 1824.

Ta-Ming chi-li 大明集禮 (Collected ceremonies of Great Ming). 1530.

Ta-Ming hui-tien. See Ming hui-tien.

Tanaka Katsumi 田中克己. "Shinsho no Shina enkai: Senkai o chūshin to shite mitaru" 清初の支那沿海：遷海を中心として見たる (The Chinese coast in the early Ch'ing, with special reference to the coastal evacuations), *Rekishigaku kenkyū*, 6.1:73–81, 6.3:83–94 (1936).

Teixeira, Manuel, "A Embaixada de Manuel de Saldanha a Pequim," *Boletim Ecclesiástico da Diocese de Macau,* No. 654 (October 1957), pp. 891–894.

Thiele, Dagmar. *Der Abschluss eines Vertrages: Diplomatie zwischen Sung- und Chin-Dynastie, 1117–1123.* Münchener Ostasiatische Studien, Band 6. Wiesbaden, Franz Steiner, 1971.

Tsao, Kai-fu. "The Rebellion of the Three Feudatories Against the Manchu Throne in China, 1673–1681: Its Setting and Significance." PhD dissertation, Columbia University, 1965.

Tso-chuan 左傳. Cited from James Legge, tr., *The Chinese Classics,* Volume 5. Reprint, Hong Kong, Hong Kong University Press, 1960.

Verbiest, Ferdinand, S.J. *Correspondance.* Ed. H. Josson and L. Willaert. Brussels, Palais des Académies, 1938.

Videira Pires, Benjamin, S.J. *Embaixada Mártir.* Macao, Centro de Informação e Turismo, 1965.

Viraphol, Sarasin. *Tribute and Profit: Sino-Siamese Trade, 1652–1853.* Cambridge, Council on East Asian Studies, Harvard University, 1977.

Vixseboxse, J. *Een Hollandsch Gezantschap naar China in de Zeventiende Eeuw (1685–1687).* Sinica Leidensa, 5. Leiden, Brill, 1946.

Walker, Richard L. *The Multi-State System of Ancient China.* Hamden, Shoe String Press, 1953.

Wang Gungwu. "The Opening of Relations Between China and Malacca, 1403-1405." In John Bastin and R. Roolvink, eds., *Malayan and Indonesian Studies: Essays Presented to Sir Richard Winstedt.* Oxford, Clarendon Press, 1964.

Wang Shih-chen 王士禎. *Ch'ih-pei ou-t'an* 池北偶談. (Chance conversations north of the pond). 2 vols. Reprinted in *Pi-chi hsiao-shuo ts'ung-shu* 筆記小說叢書 (Collection of jottings and stories). (Shanghai, 1935).

——. *Yü-yang ching-hua lu* 漁洋精華錄 (Choice works of Yü-yang). 1700; Taipei reprint, 1966.

Wang Yi-t'ung. *Official Relations between China and Japan, 1368–1549.* Harvard-Yenching Institute Studies, 9. Cambridge, 1953.

Wei Yuan 魏源. "Kuo-ch'u tung-nan ching-hai chi" 國初東南靖海記, (Record of pacification of the sea at the beginning of the dynasty). In *Hai-pin ta-shih chi* 海濱大事記 (Records of great events on the coast). *TW,* no. 213.

Wilhelm, Helmut, "From Myth to Myth: The Case of Yueh Fei's Biography," in Arthur F. Wright and Denis Twitchett, eds., *Confucian Personalities* (Stanford, Stanford University Press, 1962), pp. 146-161.

Wills, John E., Jr. "Ch'ing Relations with the Dutch, 1662–1690." PhD dissertation, Harvard University, 1967.

——. *Pepper, Guns, and Parleys: The Dutch East India Company and China, 1662–1681.* Cambridge, Harvard University Press, 1974.

——. "Early Sino-European Relations: Problems, Opportunities, and Archives," *Ch'ing-shih wen-t'i,* 3.2:50–76 (December 1974).

——. "De V.O.C. en de Chinezen in China, Taiwan, en Batavia in de 17de en de 18de Eeuw," in M.A.P. Meilink-Roelofsz, ed., *De V.O.C. in Azië,* Bussum, Unieboek, 1976, pp. 157–192.

——. "Maritime China from Wang Chih to Shih Lang: Themes in Peripheral History." In Jonathan D. Spence and John E. Wills, Jr., ed., *From Ming to Ch'ing; Conquest, Region, and Continuity in Seventeenth-Century China.* New Haven and London, Yale University Press, 1979, pp. 204–238.

——. "State Ceremony in Late Imperial China: Notes for a Framework for Discussion," *Bulletin of the Society for the Study of Chinese Religion,* 7:46–57 (Fall 1979).

——. "The Hazardous Missions of a Dominican: Victorio Riccio O.P. in Amoy, Taiwan, and Manila." In *Actes du IIe Coloque International de Sinologie, Chantilly.* Paris, Les Belles Lettres, 1980.

——. "Some Dutch Sources on the Jesuit China Mission, 1662–1687," *Archivum Historicum Societatis Iesu,* in press.

——. "Advances and Archives in Early Sino-Western Relations: An Update," *Ch'ing-shih wen-t'i,* 4.10:87–110 (December 1983).

Witsen, Nicolaes. *Noord- en Oost-Tartaryen.* Amsterdam, 1692.

Wu Ch'ang-yuan 吳長元. *Ch'en-yuan chih-lueh* 宸垣識略 (Notes from the imperial city) 1876.

Wu, Keng, Shang, K'ung ssu-wang ch'üan-chuan 吳耿尚孔四王全傳 (Complete biographies of the four princes Wu, Keng, Shang, and K'ung). *TW,* No. 241.

Wu, Silas H.L., *Passage to Power: K'ang-hsi and his Heir Apparent, 1661–1722.* Cambridge and London, Harvard University Press, 1979.

Yu T'ung 尤侗. "Wai-kuo chu-chih tz'u" 外國竹枝詞 (Ballads of foreign countries). In *Yu Hsi-t'ang chi* 尤西堂集 (Collected works of Yu Hsi-t'ang). n.d.

Yü Cheng-hsieh 俞正燮. *Kuei-ssu lei-kao* 癸巳類稿 (Classified drafts from 1833). 1833; Shanghai reprint, 1957.

Yü Wen-i 余文儀. *Hsu-hsiu T'ai-wan fu chih* 續修臺灣府志 (Revised gazetteer of Taiwan prefecture). *TW,* No. 121.

Yü, Ying-shih. *Trade and Expansion in Han China: A Study in the Structure of Sino-Barbarian Economic Relations.* Berkeley and Los Angeles, University of California Press, 1967.

Glossary

(Omitted are basic terms understood by all students of China, well-known place-names, and personal names to be found in the standard English-language biographical dictionaries.)

Chang Ch'ao-lin 張朝璘
chao 詔
Chao Ch'ang 趙昌
ch'eng-ch'en 稱臣
chieh-chien 戒見
Ch'ien-ch'ing Gate 乾清門
chin-ho 進賀
Chin Hung 金鋐
chin-kung 進貢
Chou Ch'üan-pin 周全斌
Chou Yu-te 周有德

fan-ch'en 藩臣

hai-tao 海道
ho-ch'in 和親
Ho-lan-kuo chi pa-nien i-kung, ch'i erh-nien mao-i yung-yuan t'ing-chih 荷蘭國既八年一貢,其二年貿易永遠停止
Hsi-li-t'ing 習禮亭
Hsiang-nu 降奴
Hsu Chih-chien 許之漸
Hsu Shih-ch'ang 許世昌
Hui-t'ung-kuan 會同館

k'an-ho 勘合

Li Chih-feng 李之鳳
ling 綾
lo 羅
lu 虜
Lu Ch'ung-chün 盧崇峻
Lu Hsing-tsu 盧興祖

mang-tuan 蟒緞

Shih-tzu-shuo 獅子說
shu-kuo 屬國
shui-shih t'i-tu 水師提督

T'ai-ho Gate 太和門
T'ai-ho Hall 太和殿
ti-chü-shih 提舉使

wai-ch'en 外臣
Wang Kuo-an 王國安
Wang Lai-jen 王來任
Wu Gate 午門

ying-lao 迎勞

Index

Harvard East Asian Monographs

1. Liang Fang-chung, *The Single-Whip Method of Taxation in China*
2. Harold C. Hinton, *The Grain Tribute System of China, 1845-1911*
3. Ellsworth C. Carlson, *The Kaiping Mines, 1877-1912*
4. Chao Kuo-chün, *Agrarian Policies of Mainland China: A Documentary Study, 1949-1956*
5. Edgar Snow, *Random Notes on Red China, 1936-1945*
6. Edwin George Beal, Jr., *The Origin of Likin, 1835-1864*
7. Chao Kuo-chün, *Economic Planning and Organization in Mainland China: A Documentary Study, 1949-1957*
8. John K. Fairbank, *Ch'ing Documents: An Introductory Syllabus*
9. Helen Yin and Yi-chang Yin, *Economic Statistics of Mainland China, 1949-1957*
10. Wolfgang Franke, *The Reform and Abolition of the Traditional Chinese Examination System*
11. Albert Feuerwerker and S. Cheng, *Chinese Communist Studies of Modern Chinese History*
12. C. John Stanley, *Late Ch'ing Finance: Hu Kuang-yung as an Innovator*
13. S. M. Meng, *The Tsungli Yamen: Its Organization and Functions*
14. Ssu-yü Tẹng, *Historiography of the Taiping Rebellion*
15. Chun-Jo Liu, *Controversies in Modern Chinese Intellectual History: An Analytic Bibliography of Periodical Articles, Mainly of the May Fourth and Post-May Fourth Era*
16. Edward J. M. Rhoads, *The Chinese Red Army, 1927-1963: An Annotated Bibliography*
17. Andrew J. Nathan, *A History of the China International Famine Relief Commission*
18. Frank H. H. King (ed.) and Prescott Clarke, *A Research Guide to China-Coast Newspapers, 1822-1911*
19. Ellis Joffe, *Party and Army: Professionalism and Political Control in the Chinese Officer Corps, 1949-1964*
20. Toshio G. Tsukahira, *Feudal Control in Tokugawa Japan: The Sankin Kōtai System*

21. Kwang-Ching Liu, ed., *American Missionaries in China: Papers from Harvard Seminars*
22. George Moseley, *A Sino-Soviet Cultural Frontier: The Ili Kazakh Autonomous Chou*
23. Carl F. Nathan, *Plague Prevention and Politics in Manchuria, 1910–1931*
24. Adrian Arthur Bennett, *John Fryer: The Introduction of Western Science and Technology into Nineteenth-Century China*
25. Donald J. Friedman, *The Road from Isolation: The Campaign of the American Committee for Non-Participation in Japanese Aggression, 1938–1941*
26. Edward Le Fevour, *Western Enterprise in Late Ch'ing China: A Selective Survey of Jardine, Matheson and Company's Operations, 1842–1895*
27. Charles Neuhauser, *Third World Politics: China and the Afro-Asian People's Solidarity Organization, 1957–1967*
28. Kungtu C. Sun, assisted by Ralph W. Huenemann, *The Economic Development of Manchuria in the First Half of the Twentieth Century*
29. Shahid Javed Burki, *A Study of Chinese Communes, 1965*
30. John Carter Vincent, *The Extraterritorial System in China: Final Phase*
31. Madeleine Chi, *China Diplomacy, 1914–1918*
32. Clifton Jackson Phillips, *Protestant America and the Pagan World: The First Half Century of the American Board of Commissioners for Foreign Missions, 1810–1860*
33. James Pusey, *Wu Han: Attacking the Present through the Past*
34. Ying-wan Cheng, *Postal Communication in China and Its Modernization, 1860–1896*
35. Tuvia Blumenthal, *Saving in Postwar Japan*
36. Peter Frost, *The Bakumatsu Currency Crisis*
37. Stephen C. Lockwood, *Augustine Heard and Company, 1858–1862*
38. Robert R. Campbell, *James Duncan Campbell: A Memoir by His Son*
39. Jerome Alan Cohen, ed., *The Dynamics of China's Foreign Relations*
40. V. V. Vishnyakova-Akimova, *Two Years in Revolutionary China, 1925–1927*, tr. Steven I. Levine
41. Meron Medzini, *French Policy in Japan during the Closing Years of the Tokugawa Regime*
42. *The Cultural Revolution in the Provinces*
43. Sidney A. Forsythe, *An American Missionary Community in China, 1895–1905*
44. Benjamin I. Schwartz, ed., *Reflections on the May Fourth Movement: A Symposium*
45. Ching Young Choe, *The Rule of the Taewŏn'gun, 1864–1873: Restoration in Yi Korea*

46. W. P. J. Hall, *A Bibliographical Guide to Japanese Research on the Chinese Economy, 1958–1970*
47. Jack J. Gerson, *Horatio Nelson Lay and Sino-British Relations, 1854–1864*
48. Paul Richard Bohr, *Famine and the Missionary: Timothy Richard as Relief Administrator and Advocate of National Reform*
49. Endymion Wilkinson, *The History of Imperial China: A Research Guide*
50. Britten Dean, *China and Great Britain: The Diplomacy of Commerical Relations, 1860–1864*
51. Ellsworth C. Carlson, *The Foochow Missionaries, 1847–1880*
52. Yeh-chien Wang, *An Estimate of the Land-Tax Collection in China, 1753 and 1908*
53. Richard M. Pfeffer, *Understanding Business Contracts in China, 1949–1963*
54. Han-sheng Chuan and Richard Kraus, *Mid-Ch'ing Rice Markets and Trade, An Essay in Price History*
55. Ranbir Vohra, *Lao She and the Chinese Revolution*
56. Liang-lin Hsiao, *China's Foreign Trade Statistics, 1864–1949*
57. Lee-hsia Hsu Ting, *Government Control of the Press in Modern China, 1900–1949*
58. Edward W. Wagner, *The Literati Purges: Political Conflict in Early Yi Korea*
59. Joungwon A. Kim, *Divided Korea: The Politics of Development, 1945–1972*
60. Noriko Kamachi, John K. Fairbank, and Chūzō Ichiko, *Japanese Studies of Modern China Since 1953: A Bibliographical Guide to Historical and Social-Science Research on the Nineteenth and Twentieth Centuries, Supplementary Volume for 1953–1969*
61. Donald A. Gibbs and Yun-chen Li, *A Bibliography of Studies and Translations of Modern Chinese Literature, 1918–1942*
62. Robert H. Silin, *Leadership and Values: The Organization of Large-Scale Taiwanese Enterprises*
63. David Pong, *A Critical Guide to the Kwangtung Provincial Archives Deposited at the Public Record Office of London*
64. Fred W. Drake, *China Charts the World: Hsu Chi-yü and His Geography of 1848*
65. William A. Brown and Urgunge Onon, translators and annotators, *History of the Mongolian People's Republic*
66. Edward L. Farmer, *Early Ming Government: The Evolution of Dual Capitals*
67. Ralph C. Croizier, *Koxinga and Chinese Nationalism: History, Myth, and the Hero*
68. William J. Tyler, tr., *The Psychological World of Natsumi Sōseki*, by Doi Takeo

69. Eric Widmer, *The Russian Ecclesiastical Mission in Peking during the Eighteenth Century*
70. Charlton M. Lewis, *Prologue to the Chinese Revolution: The Transformation of Ideas and Institutions in Hunan Province, 1891–1907*
71. Preston Torbert, *The Ch'ing Imperial Household Department: A Study of its Organization and Principal Functions, 1662–1796*
72. Paul A. Cohen and John E. Schrecker, eds., *Reform in Nineteenth-Century China*
73. Jon Sigurdson, *Rural Industrialization in China*
74. Kang Chao, *The Development of Cotton Textile Production in China*
75. Valentin Rabe, *The Home Base of American China Missions, 1880–1920*
76. Sarasin Viraphol, *Tribute and Profit: Sino-Siamese Trade, 1652–1853*
77. Ch'i-ch'ing Hsiao, *The Military Establishment of the Yuan Dynasty*
78. Meishi Tsai, *Contemporary Chinese Novels and Short Stories, 1949–1974: An Annotated Bibliography*
79. Wellington K. K. Chan, *Merchants, Mandarins, and Modern Enterprise in Late Ch'ing China*
80. Endymion Wilkinson, *Landlord and Labor in Late Imperial China: Case Studies from Shandong by Jing Su and Luo Lun*
81. Barry Keenan, *The Dewey Experiment in China: Educational Reform and Political Power in the Early Republic*
82. George A. Hayden, *Crime and Punishment in Medieval Chinese Drama: Three Judge Pao Plays*
83. Sang-Chul Suh, *Growth and Structural Changes in the Korean Economy, 1910–1940*
84. J. W. Dower, *Empire and Aftermath: Yoshida Shigeru and the Japanese Experience, 1878–1954*
85. Martin Collcutt, *Five Mountains: The Rinzai Zen Monastic Institution in Medieval Japan*

STUDIES IN THE MODERNIZATION OF THE REPUBLIC OF KOREA: 1945–1975

86. Kwang Suk Kim and Michael Roemer, *Growth and Structural Transformation*
87. Anne O. Krueger, *The Developmental Role of the Foreign Sector and Aid*
88. Edwin S. Mills and Byung-Nak Song, *Urbanization and Urban Problems*
89. Sung Hwan Ban, Pal Yong Moon, and Dwight H. Perkins, *Rural Development*

90. Noel F. McGinn, Donald R. Snodgrass, Yung Bong Kim, Shin-Bok Kim, and Quee-Young Kim, *Education and Development in Korea*
91. Leroy P. Jones and Il SaKong, *Government, Business and Entrepreneurship in Economic Development: The Korean Case*
92. Edward S. Mason, Dwight H. Perkins, Kwang Suk Kim, David C. Cole, Mahn Je Kim, et al., *The Economic and Social Modernization of the Republic of Korea*
93. Robert Repetto, Tai Hwan Kwon, Son-Ung Kim, Dae Young Kim, John E. Sloboda, and Peter J. Donaldson, *Economic Development, Population Policy, and Demographic Transition in the Republic of Korea*
106. David C. Cole and Yung Chul Park, *Financial Development in Korea, 1945-1978*

94. Parks M. Coble, *The Shanghai Capitalists and the Nationalist Government, 1927-1937*
95. Noriko Kamachi, *Reform in China: Huang Tsun-hsien and the Japanese Model*
96. Richard Wich, *Sino-Soviet Crisis Politics: A Study of Political Change and Communication*
97. Lillian M. Li, *China's Silk Trade: Traditional Industry in the Modern World, 1842-1937*
98. R. David Arkush, *Fei Xiaotong and Sociology in Revolutionary China*
99. Kenneth Alan Grossberg, *Japan's Renaissance: The Politics of the Muromachi Bakufu*
100. James Reeve Pusey, *China and Charles Darwin*
101. Hoyt Cleveland Tillman, *Utilitarian Confucianism: Ch'en Liang's Challenge to Chu Hsi*
102. Thomas A. Stanley, *Ōsugi Sakae, Anarchist in Taishō Japan: The Creativity of the Ego*
103. Jonathan K. Ocko, *Bureaucratic Reform in Provincial China: Ting Jih-ch'ang in Restoration Kiangsu, 1867-1870*
104. James Reed, *The Missionary Mind and American East Asia Policy, 1911-1915*
105. Neil L. Waters, *Japan's Local Pragmatists: The Transition from Bakumatsu to Meiji in the Kawasaki Region*
108. William D. Wray, *Mitsubishi and the N.Y.K., 1870-1914: Business Strategy in the Japanese Shipping Industry*
109. Ralph William Huenemann, *The Dragon and the Iron Horse: The Economics of Railroads in China, 1876-1937*
110. Benjamin A. Elman, *From Philosophy to Philology: Intellectual and Social Aspects of Change in Late Imperial China*

111. Jane Kate Leonard, *Wei Yuan and China's Rediscovery of the Maritime World*
112. Luke S. K. Kwong, *A Mosaic of the Hundred Days: Personalities, Politics, and Ideas of 1898*
113. John E. Wills, Jr., *Embassies and Illusions: Dutch and Portuguese Envoys to K'ang-hsi, 1666–1687*
114. Joshua A. Fogel, *Politics and Sinology: The Case of Naitō Konan (1866–1934)*